IF MAPS COULD SPEAK

To Eilís

IF MAPS COULD SPEAK

RICHARD KIRWAN

LONDUBH BOOKS

First published in 2010 by

Londubh Books

18 Casimir Avenue, Harold's Cross, Dublin 6w, Ireland

www.londubh.ie

1 3 5 4 2

Cover by bluett;

The cover image incorporates a detail from a manuscript map of Waterford 1841, cour-
tesy of the National Archives, and an engraving of Poolbeg Lighthouse by permission of
Ordnance Survey Ireland.

Origination by Londubh Books

Printed in Ireland by ColourBooks, Baldoyle Industrial Estate, Dublin 13

ISBN: 978-1-907535-093

Acknowledgements

I would like to acknowledge with gratitude my publisher, Jo O'Donoghue of Londubh Books, who suggested that I should write this book, and the support I received from her throughout the process.

I am particularly grateful to Eilís, my wife, who patiently read every draft and gave me advice and suggestions on the content and structure of the book.

I wish to extend my thanks to Abbot Patrick Hederman of Glenstal Abbey for his ongoing encouragement and for providing the foreword. He and the monks of Glenstal provided me with a beautiful place in which to write part of this book. I am grateful to Fanny Howe, the American poet, whom I befriended at Glenstal and who read the manuscript as it developed and offered much-appreciated critical advice. Thanks to the many friends I made at Glenstal, including Simon Sleeman and Sue Moon, and sincerest thanks to my writers' group at Glenstal: Anne O'Connor, P.J. McAuliffe, Eva O'Callaghan, Denise Collins and Nuala Keher. Thanks also to Dennis Hamley for his editorial advice.

I wish to express my thanks to Geraldine Ruane, CEO of Ordnance Survey Ireland, for her support, and to Andy McGill, Brian Madden, Maurice Kavanagh, Mairéad Harrington and Peter Hallahan, who helped me with the selection of the images. Many thanks to Paul Ferguson of the Map Library in Trinity College for all his help and to the staff of the National Archives and the RIA. Michelle Jackson provided me with the drawing of the Watch Tower, for which I am very grateful. Thanks to John Danaher,

Malachy McVeigh and Michael Harford for prodding my memory and to Dominic McNamara of St Patrick's College Maynooth for generously sharing his knowledge.

The names of some people referred to in this book have been changed to protect their privacy.

CONTENTS

Foreword by Mark Patrick Hederman

Clifford Geertz, the anthropologist, advises us that when we are exploring new territory we should bring with us a poet and a cartographer. But in Ireland the cartographer was left out, at least until this essential book arrived, and the poets have had it all their own way.

In 'The Solitude of Alexander Selkirk' the eighteenth-century English poet, William Cowper, wrote about the original Robinson Crusoe on his desert island:

> I am monarch of all I survey,
> My right there is none to dispute;
> From the centre all round to the sea
> I am lord of the fowl and the brute.

The monarchical ring of the word 'survey', when coupled with the word 'ordnance', meaning 'the branch of an armed force that procures, maintains, and issues weapons, ammunition, and combat vehicles', caused hackles to rise, especially in places where ordinary people were regarded as little better than brutes. For many people in Ireland, Ordnance Survey was synonymous with British colonisation.

Richard Kirwan sees it differently. If maps could speak, he suggests, they would do so neither in Irish nor in English. They speak a much more fundamental language, the language of geography before any languages were spoken.

So what is a map? The word 'map' comes from the Latin for a handkerchief or a piece of cloth. A map is the way we represent the planet on which we live on a tiny two-dimensional surface: the world sketched out on my handkerchief.

Experts tell us that the earth came into being some 4600 million years ago. There were no such things as maps at that time. The most successful animals to inhabit the planet for over two million years, the dinosaurs, had no maps. Maps are essentially a human phenomenon. Human beings need maps to understand the landscape that surrounds them. Our minds translate the space around us into geometry, which comes from Greek and means 'measurement of the world'. Distance, length and volume are the basic tools we use for organising the multitude of shapes and sizes that nature shoots up at us like a firework display at every moment. Trigonometry, which helps us with our map-making, comes from the Greek for 'measurement by triangles': hence the 'triangulation programme' which formed the basis of what is beautifully described in this book.

Why was Ireland the best mapped country in the world at the start of the twentieth century? Not because the country was under British rule or because it was mapped by the British army but because it happens to have had at its disposal a number of gifted and extraordinary mapomaniacs who were in the right place at the right time to undertake this work. They didn't care who was employing them: their passion was for making maps as accurately as possible. This book tells the story of some of these individual geniuses.

Brian Friel's play *Translations* (1980) poetically considers the first Ordnance Survey mapping of Ireland as a metaphor for the colonial relationship between Ireland and England. Because map-making renames the landscape and translates its placenames from Irish into English it assumes the role of colonial usurper in the play.

Richard Kirwan shows that map-making in itself is innocent of all such colonial tendencies. Placenames, whether in English or in Irish, are later impositions on the landscape, like a branding of the skin. The art of map-making concerns itself with the skin itself and its ways of identifying, measuring and classifying are symbolic codes that are antecedent to the poetic function of naming and describing.

A map is meant to be an accurate objective diagram of a place. Those who commissioned the maps may have had ulterior motives of a political or economic nature but those who undertook the Ordnance Survey work, as described in this fascinating book, were dedicated cartographers whose only obsession was with the business of map-making as an artform in itself. They gave their lives to the undervalued and often misunderstood task of articulating in detail the contours of the country in which we live.

No one is better qualified to give us such an account of how maps can describe both the landscape of our country and the ways in which we have progressed over nearly two centuries. Richard Kirwan combines a unique passion for the local geography of countryside and townland with a lifetime spent assiduously mapping out this territory. If maps could choose they would certainly want to use him as their spokesperson. In this exciting book he has admirably honoured the vocation of map-making. It is both a personal story and a detailed and unbiased account of the history of mapping in this country.

PROLOGUE

It was love at first sight. I can have been no more than six years old when I first discovered a very small, very old map of Waterford hanging in a shop window. Instantaneously, it printed an indelible image on my mind. This fan-shaped city, where the roads like long fingers led to the unknown of the surrounding countryside, could not be mistaken for any other. This had been its shape since Viking times.

I don't know why a map should have held such a fascination for me at that age. Perhaps it was because books were a rare commodity in our house and I had little to stir my imagination. There were only three that I can remember: *Treasure Island*, *Robinson Crusoe* and *Black Beauty*. I detested *Black Beauty*. *Treasure Island* had its island and so did *Robinson Crusoe*. Even better, Crusoe's was deserted. I could play in my imagination with the maps of these islands, visualise their shapes and sizes, travel their endless rocky pathways and explore places no one had ever been. Some paths led to scary places and dead ends where strange wild animals lived. Others brought me to treasure and freedom.

It was strange that of all the fairy tales I heard on our kitchen radio Hans Christian Andersen's 'The Little Match Girl' was the one that should haunt me for ever. I imagined the little girl to be in Waterford at Christmas time. I could see her, even travel with her, through the streets. I felt her loneliness, her cold and hunger. I looked through her eyes into the windows of the shops in Michael Street and longed for a tiny piece of what we saw. I felt her pain when passers-by rejected our pleas and would not buy a match. Her cry tore me apart when she looked through the windows of the big houses on the Mall and wished for a little plate of plum pudding from the revellers' dining tables. The images became a permanent part of my mental map of the city.

When I was a small child, my father and grandfather took me on journeys of discovery throughout the city. I inscribed in my mind a sketch of Waterford that mesmerised me. It was the 1950s and to most people it was a dull city from whose quayside the Great Western carried men and women to England in search of work.

For me, it was a city full of old buildings: ancient castles and towers along the original city walls; old houses, some like those along the Mall, tall, noble and hauntingly beautiful; others hidden in laneways, frighteningly dark or almost derelict. The city skyline fascinated me; it was a map in its own right. It was dominated by a single spire, that of the Protestant cathedral. The belfries of the numerous churches – the Friary, the Dominican, Ballybricken Church, the ruined Blackfriars and the French Church – stretched upwards to show themselves above the grey rooflines. All the shop fronts, like Hearne's and Bell's, were unique and most of them were old. There was a mixture of wide and narrow streets and laneways, some with steep steps, leading to the city centre. Everywhere was full of the reminders of other times; the new at the edges of the city, not yet endowed with any reminiscences, was just emerging.

I daydreamed whenever I walked by the old river that flowed endlessly to Dunmore East and the sea, oblivious to the hustle and bustle on the Quay at its edge, unimpressed by the elegance of the ships that sailed its waters. Old ships were still coming and going: black steamships, even the occasional sailing ship. Other ships, their journeys finished for ever, like the abandoned ship constructed from concrete moored just upstream from the bridge, were slumped by the quayside, slowly and forlornly listing, slipping into the accommodating, silky mud deposited by the millennia of flowing water.

It should have been obvious to me from that early childhood that making maps would come naturally to me. It was no more than a continuation of childhood dreams, an extension of what my father and grandfather had taught me on our journeys of discovery, a love of landscape. They had guided me to be an observer of town and country, a maker of mental maps. But I never saw it that way; I thought it was just part of the magic of my childhood world.

As a child I was shy and sensitive but single-minded, conspicuous with my cow's-lick and hair parting on the right-hand side. I got on well with my friends on the street and in school and spent a lot of my time enjoying the outdoor activities that boys engaged in; playing cowboys and Indians, marbles, hide and seek; exploring all that was exciting in our surroundings; taking the long routes home from school; even annoying elders with childish pranks to entice a chase. Games like hurling and football became important to me as I grew up. But in my shyness I found it difficult to approach people I did not know well or to ask questions about things I did not understand or was curious about. However, I was a good student and what I failed to find out by asking questions I learned in other ways. I possessed a natural observational inquisitiveness which I used in a determined way to figure out challenges that interested me and to answer questions that engaged my mind.

Perhaps that was why I so intensely identified with the intimacies and tapestry of the city during my formative years. I felt that I was not just an onlooker, a passer-by, but that I had always been an integral part of it and its people, especially those who worked in the factories and shops and lived in the streets where I lived, walked, played and attended school. For me the city and its people were one, each enhancing the character of the other.

The real characters of all the people I knew, no matter how tenuous these acquaintances, stepped into the world only when I saw them in their homes, streets or places of work. Places had little meaning without their people. My grandfather and father showed me different aspects of this city and its people; they were my educators, my extended self. I learned as I walked with them through the city centre, past the factories and through Ballybricken, home of Waterford's ceremonial occasions.

In Waterford I enthusiastically soaked up the knowledge of place and people my forebears imparted to me, getting to know the patterned fabric, the texture and the uniqueness of the city through sight, smell, sound and the spoken word. As my knowledge of the city grew I noticed the preoccupation people had with boundaries, real and imaginary. The solid demarcation of buildings and property divisions – walls, iron railings, hedges, drainpipes, colour and design – staked out their limits. Territorial divisions were mostly invisible, often virtual; the electoral ward boundaries running along the centre of streets, imaginary lines where one street ended and the other began, and meandering, invisible but safeguarded boundaries of districts in the city where different cultures met: the rich and the poor, the new and the old.

I came to know this city so well that I could call the detailed image of any part of it to my mind instantly. It was as if an impression of the city map was etched permanently within me, a map through which I could roam at will. But this representation in my mind was not like the old folded paper map from the 1930s which I kept on my bedroom shelf and which I spent endless days exploring. This map depicted the streets and the stark outline of blocks of buildings. It showed the street names in cold and solid black letters. The picture I had painted in my mind was altogether different; it was multidimensional, rich with layers of living detail and images within images. Each street held intense and individual evocations; complete pictures of all the buildings, their age and their character: happy, sad, young, aged, wealthy or poor.

The buildings themselves contained further clues to the characters of the people who lived there: a door with its proud glossy colour or dull flaking paint; a letterbox, knocker, knob or keyhole of shining brass or dull and green from neglect.

Windows with glass shining and transparent or dirty, almost opaque with grime and dust, curtains half-drawn or fully open, stiff with neat folds or limp from lack of care, pots of geraniums red and luscious or bedraggled and struggling from drought, ornaments crisp and neat or fallen, chipped and abandoned.

I could sense an individual aura attached to each street and its buildings, an aura that became part of my mental map. In some places it was light and happy like those bright, almost translucent, weightless wisps of cloud that appear from nowhere, drift and float and continuously alter shape in an otherwise blue sky. In other streets the aura was dark and dreary, slumping around buildings like grey, leaden, rain-filled clouds waiting to dump their gloom. And in places it was neutral, ready to assume the changing moods of the days or the inhabitants. Even more pronounced were the recollections from my relatives' lives, the sensations from my father's stories and most of all my own feelings, born from all I experienced through sight, smell or sound, anchored forever to the sites of their creation.

The surrounding country and local seaside resorts had their own distinctive allure for me. The people who lived there, their interests and lifestyles, were different from their city neighbours; nature saw to that. Their faces and hands were furrowed and a deeper colour, gifts the sun, rain and wind had fashioned and shaped from endless days of exposure. They were in general more accepting and lived life at an easier pace, more in tune with nature's unrushed time-piece.

My association with every detail of the countryside was not as intense as the one I had with the city; in places there was a more general relationship, a tuning in to the moods of the more open landscape, the animals and the birds, a silent understanding of its personal history. With other places like Tramore and Woodstown, I had a more intimate association – with deep images, memories and emotions formed by long, hot summer holidays – as I had with places less obvious, like the groves of trees where the crows cawed loudest, the mushroom fields, my relatives' farmyard.

I finally got to know the relationships between all these places within the bigger landscape from a 1906 coloured map of the city hinterland that also sat on my bedroom shelf. Unlike the lifeless monochrome map of the city this one had warmth and substance about it: hachured depiction of relief, green forests, brown roads, blue waters. Even the texture of the linen it was printed on made it more friendly. This rural landscape was also etched on my mind, just like a map engraved in copper. But, like my mental map of the city, it was enriched by layers of images, sounds, smells and emotions.

Despite my knowledge, the city and country always threw up

surprises. In the city a rare new shop might, unknown to me, have opened for business, or a long-established one have closed suddenly. In the countryside, every new winding road or track I cycled along revealed something new, something more to sketch on my mental map. Their twists and turns often showed something so unexpected that it reminded me of the unknown that lay ahead. The unknown was not always made of matter, a village suddenly discovered or an underground river. This unknown would stay unknowable, incomprehensible; it was my destiny.

I spent eighteen years growing up in that city. I departed as suddenly as I had arrived, when I drove away in September of 1966 on my way to a life full of promise. I forsook my mental map and all it contained. There were wonderful memories in it but there were some on which it was too painful to dwell. I was determined to fulfil a dream to be a civil engineer, to build roads and bridges over which life would take me.

But I might have known that those old maps would have the last word.

1

GO WHERE THERE IS NO PATH

When I was first introduced to the Ordnance Survey in 1970 I was given two pieces of information by the then Assistant Director, Gerry Madden. The first was that the Ordnance Survey in Ireland began in 1824 under the Board of Ordnance, of which the Duke of Wellington was in charge. It derived the name 'Ordnance' from there. The second was that Ireland was the best-mapped country in the world by the time it gained its independence in 1922.

It was a strange name for a mapping organisation but it made sense when he told me. It was a surprise to hear that the Duke of Wellington had a hand in starting it but it probably needed someone of his stature and influence to get such a monumental work started in the Ireland of that time. It meant that there were two monuments to him in the Phoenix Park. One was visible: the tall stone column at the entrance to the Park which celebrated his military victories and was at the time the tallest monument in Europe. The other, bearing no visible reference to him, was hidden among the trees at the other end of the Park. This was his unsung masterpiece. It was at least equal to his other monument.

In the early 1970s little had been written on the history of the Ordnance Survey in Ireland, apart from pieces in books where events in the Survey of Great Britain intertwined with those of the Irish Survey during the nineteenth century.

It was left to a professor of geography in Trinity College to unfold the intricacies of that history. John Andrews published *A Paper Landscape* in 1975, after years of exhaustive research. I often met this pale-faced man with thick-rimmed glasses emerging from the manuscript stores or ascending the stairs from the rat-friendly basement of Mountjoy House. I was told that he was a researcher but nobody seemed to take any notice of him. He was a serious man, engrossed in his work, who always appeared to be mentally juggling with some complicated puzzle or mining the depths of his mind for lost scraps of information. John spent every minute of his available time searching the records in Ordnance Survey, the Royal Irish Academy, the National Archives and elsewhere, seeking information to piece together the story of map-making in Ireland.

The Ordnance Survey, a military organisation, was set up in Britain in 1791 to map the south of England and parts of Scotland. Good topographic maps were needed by the military to plan defences against a possible attack from France by Napoleon.

The Ordnance Survey in Ireland was established for a different reason. New maps were needed so that an equitable taxation system could be set up and implemented. The existing taxation system was called 'the county cess'. It was used to fund public works such as the construction and maintenance of roads and bridges and to pay local officials. This tax was recognised as being inequitable. It could never have been otherwise as the system was based on the rateable valuation of each townland. The boundaries of the townlands were known but the acreages attributed to most of them were not reliable. It was therefore impossible to apportion the tax fairly among landowners. In some counties, such as County Clare, maps made by Lord Strafford during the plantations in the 1630s were used as a basis for taxation. The original maps were no longer in existence, so they depended on inferior and inadequate copies. Other counties used William Petty's *Down Survey* from later in the seventeenth century.

Since the early 1800s, there had been constant agitation and political manoeuvrings by the land-owning gentry about the land taxation system in Ireland. The British Parliament began to take notice, so much so that four parliamentary committees heard submissions on this vexed question between 1815 and 1824. The Spring-Rice committee of 1824 recommended that entirely new maps of Ireland should be made and that the Ordnance Survey of Great Britain should extend its remit into Ireland. The Spring-Rice report was signed on 21 June 1824 and Ordnance Survey in Ireland came into being one day later, on 22 June 1824.

Major Thomas Colby had been appointed to lead the British Ordnance Survey in 1820. His duties were extended to include the new mapping of Ireland. Colby was born in 1774 into a military family. His father had been a Royal Marine. Colby was commissioned a lieutenant in the Royal Engineers in 1801 at the age of seventeen and immediately posted to the Ordnance Survey in Britain. His uncle, General Haddon, was undoubtedly instrumental in his posting. General William Mudge, the Director of Ordnance Survey, was pleased with his new officer and wrote, 'I find him on examination, well grounded in the rudiments of mathematics, and in other respects perfectly calculated to be employed in his business.'

Colby spent many of his early years working on the triangulation programme – dividing the country into a series of triangles so that it could be accurately mapped – in the Highlands of Scotland. His enthusiasm and single-minded devotion to his work were evident. He had lost his left hand and his skull had been fractured shortly after he was commissioned, when a loaded pistol exploded, but he did not let his handicap interfere with his sense of duty or the intensity with which he carried out his work.

Lieutenant Dawson, a young officer working with him, kept an account of the 1819 triangulation campaign which Colby led. It began in June in Huntly, north of Aberdeen, and concluded at the end of September. Colby travelled from London by the mail-coach, a journey which took up to five days and nights and on which he took only a short rest in Edinburgh. Dawson tells us that '…neither rain nor snow, nor any degree of severity in the weather, would induce him to take an inside seat or to tie a shawl round his throat'. He ate very little on his journey – only meat and bread with tea or a glass of beer. He immediately joined the survey teams and partly climbed Corrie Habbie Mountain, where he slept in his clothes on a bundle of tent-linings in a marquee, but not before he helped the men to erect their tents. The following day the team climbed to the

summit of the two-thousand-foot mountain where Colby helped
with setting up the survey station and shelter for the men. After that
he carried out observations with the theodolite.

During that summer of 1819 Colby undertook two major
'station-hunts', expeditions in search of new observation points.
He walked five hundred and thirteen miles in twenty-two days on
one expedition and five hundred and eighty-six miles over a similar
period on the other. These were no ordinary journeys. He scaled the
mountain peaks and ran through the valleys, taking direct routes
and ignoring the easier ways along the roads. Dawson accompanied
him on the second expedition from Corrie Habbie, which began
on 23 July. 'The party set off on a steeple-chase, running down the
mountain-side at full speed, over Cromdale, a mountain about
the same height as Corrie Habbie, crossing several beautiful glens,
wading the streams which flowed through them, and regardless
of all difficulties that were not absolutely insurmountable.' If he
encountered a road, Colby allowed the weary members of the party
to travel on it but only in the spirit of rivalry to see which team
would reach its destination first. Dawson was exhausted on the
second day but, as he was an officer, Colby insisted that he travel
across the mountains with him and not by road with the more
exhausted men. That day they set out at nine in the morning and
arrived at their final destination close to midnight. That was not
untypical. Later Dawson praised Colby for insisting that he continue
on that day for it hardened him for the remainder of the expedition
and he never suffered from exhaustion again.

Good food and comfortable beds would have been a welcome
reward at the end of an arduous journey but sometimes they did
not materialise. In Cluny, the public-house, the party's abode for
a night, was a mud hovel with two beds built into the walls. Two
drovers occupied them and each of Colby's team had to make do
with 'three or four wooden chairs, placed as evenly as the earthen
floor would permit, and with our knapsacks for pillows and our
short walking-cloaks for covering'. That was after a putrid tail-end of
salmon had been served for dinner. But there were compensations.
At times, the station-hunt party was entertained by local landlords,

something which was most welcome when mist prevented work from continuing. Dawson remarked, when he was stationed at Letterew on the northern shores of Loch Maree, 'that had there been a prevalence of misty weather, or had our object been pleasure, we might, I believe, have remained, or been passed on in a similar manner from one hospitable mansion to another'.

Colby took good care of his men: as was the custom at the end of an observation campaign, he gave them the keys of the provision chests to prepare a farewell feast of their liking. Traditionally, an enormous plum pudding was made and this time it weighed almost one hundred pounds and took twenty-four hours to cook.

Colby travelled throughout Ireland in like manner, searching out suitable mountain-tops for the primary triangulation. It was an extraordinary undertaking to travel in the remotest parts of the country, when roads were poor or non-existent and when the available maps were unreliable or grossly inaccurate. But despite the difficulties and demands of the job he was very aware of the need to respect local customs. While in Scotland he was obliged to ask Dawson to stop whistling on Sunday as it offended the local observance of the Sabbath and the landlord feared that some misfortune might befall him. In Ireland he was acutely aware of sentiment towards the British, particularly the military and police. He forbade his sappers to carry arms or to be part of any normal military or police activity.

Ordnance maps in Britain, made to deter invasion, were compiled at a scale of one inch to the mile. The extension of Colby's duties to Ireland presented a different challenge. As maps in Ireland were required for taxation of land they needed to be far more detailed than their British counterparts. The Spring-Rice committee had decided on a scale of six inches to the mile. It wanted the topographic information, such as roads and rivers, to be mapped, but more importantly it wanted the townland boundaries shown accurately. There were over sixty thousand of these.

This was a far greater undertaking than the one-inch mapping of Britain. There was no precedent for it for an entire country. There had been a six-inch map of Kent made by the British Ordnance

Survey which gave few clues as to how an entire country might be mapped. It was expected that Ireland would be completely mapped in about five years and that the project would cost about £300,000. Nobody in their wildest dreams thought that it would take twenty-five years and over two thousand men to complete it. The country would surely never have been mapped to such an extent if this had been foreseen.

Many of the private surveyors working in Ireland must have assumed that they would find work with the new Ordnance Survey. After all, they had knowledge of the native people and the landscape. There was a growing number of competent cartographers, surveyors and engineers already mapping parts of the country. William Edgeworth had made a scientifically-based map of Longford. He had even travelled to Munich to have a very precise theodolite made by a renowned instrument-maker and while in Europe had studied the mapping methods used in Austria.

The Bogs Commission had employed many engineers, including three from Scotland – Richard Griffith, Alexander Nimmo and William Bald – to work on maps of the bogs of Ireland between 1809 and 1813. They continued to make maps when the bogs-mapping programme was ended. Richard Griffith worked with William Edgeworth to complete a map of Roscommon. Alexander Nimmo continued surveying roads and making charts for the Irish Fishery Board. William Bald made an extensive map of Mayo. There were others such as David Aher, who, having finished working on the bog maps, made a trigonometric map of County Kilkenny.

None of these men was taken on by the Ordnance Survey. There were no civil surveyors employed in the early days. Colby wanted Ireland to be mapped by the military only. In any event, he did not trust civilians. He had been unimpressed by the six-inch map of Kent made by civil surveyors. He ignored the surveyors' pleas and applications for employment.

The Lord Lieutenant of Ireland, Marquis Wellesley, did not want them either. He expressed his views strongly to his brother, the Duke of Wellington, when he wrote: 'The proposed survey cannot be executed by Irish engineers and Irish agents of any description.

Neither science, nor skill, nor diligence, nor discipline, nor integrity, sufficient for such a work can be found in Ireland.'

It was an unfair comment on the work of private cartographers and surveyors such as Edgeworth, Nimmo and Griffith. But the Duke of Wellington shared the opinion of the Lord Lieutenant: 'I positively refused to employ any surveyor in Ireland on this service.'

Thomas Colby brought officers from the corps of Royal Engineers, with whom he had worked in Scotland, to supervise the work. The men were drawn from the corps of Royal Sappers and Miners and were generally known as sappers. The most famous of his officers was Thomas Drummond, a man of great scientific ability, so much so that the poet Wordsworth referred to him as 'Drummond of calculating celebrity'. He worked with Colby in the early years to help him to perfect the measuring bars. It was he who invented the limelight.

Thomas Colby, like Drummond, was a man of science, meticulous and particular in everything he did. But he was also a man of very strong character and did not easily accept criticism of his methods of mapping. He had serious disagreements with his senior officers, including Thomas Drummond and Thomas Larcom, to such an extent that Drummond left the Ordnance Survey in 1831. Later he became Under-Secretary for Ireland. Yet Colby had a very human side. He got on well with the workers and insisted that the children of the sappers coming from England be given good schooling.

The setting up of the Ordnance Survey in Ireland, together with the refusal to employ private surveyors, was a watershed in Irish map-making. The Spring-Rice committee recommended that this new mapping survey would supersede all existing topographic maps, no matter how good they were. Not even Edgeworth's and Griffith's Roscommon map would remain official. From now on the great Petty maps of the seventeenth century and the others would be confined to the record books and become the preserve of historians, used only to provide clues to the past and records of what had been: what the topography and settlement patterns might have been; how names might have been spelled; what the landed estates of the gentry might have looked like, as the poor and their botháns never featured in the depiction of the landscape. The old maps would also become the repository of dreams and folklore.

The recommendation of the Spring-Rice committee effectively nationalised mapping in Ireland and spelled the end of a private industry. It was probably the first nationalisation in Ireland. Ordnance Survey maps became the official documents of the state and there was no need for any others. But the establishment of the Ordnance Survey brought with it another fundamental change. Colby, and his assistant, Thomas Larcom, introduced the processes of the industrial revolution to map-making: an assembly-line system where every man had his particular speciality. Each person carried out his piece of the map-making operation – field-survey, drawing, area-measurement or name-placement.

This was the end for the holistic surveyor-cum-cartographer, a true artist possessing the skills to make the entire map and put his own stamp and style on the depiction of the landscape. The new ways led to efficiencies and standardisation. Fortunately the style of map-making adopted ensured that it retained artistic merit and was not just a series of lines. In time a picture of the lifestyle of those who lived in the townlands of Ireland would be evoked by each map created.

The methods introduced in the first half of the nineteenth-century were still in operation when I came to the Ordnance Survey. There was still an office for each process of map-making and quite possibly the same processes being carried out in them. The drawing office was a relic of the past, as was the area-measurement office which looked just as it might have done one hundred and fifty years earlier. The typing office still had trays of lead letters with which to compose names. There was only one big change; there was less work being carried out and fewer people in the offices.

Thomas Colby began the six-inch mapping project in the north of the country and intended to work his way south. But his progress was not as fast as he had expected. There was the unpredictable Irish weather to struggle with. Sometimes his men waited for months on mountain-tops for the rain and mist to clear. There were logistical matters: recruitment and training of new surveyors, perfecting the instrumentation needed for such a scientific survey, instruments such as measuring bars and the limelight. There was an initial scarcity of theodolites.

There were other reasons for the slow progress, such as the poor quality of some of the surveyors and an apparent conspiracy between the wind and the theodolites. It was reported that 'when the survey first commenced the wind was continually blowing down instruments'. Thomas Colby responded to this damage by declaring that 'if the wind blew down any more the men should pay for them.' Miraculously, the wind ceased its sabotage of the theodolites!

The native population did not always cooperate with this invasion of surveyors. People in some parts of the country were suspicious that this new mapping might not be for their good. Sighting poles erected on mountain-tops in Donegal and Wicklow were removed. In Wexford and Bantry, surveyors had to be given police protection. Local tenants, who were often at the mercy of landlords, feared that the new mapping would have the effect of increasing rents. But matters were different in County Clare. The *Dublin Evening Post* reported in 1828 that the inhabitants of Glenomara had helped engineers to build a trigonometric station. Evidently, many of them climbed their local mountain in celebratory style with flutes, pipes, fiddles and young women bearing laurel leaves. They named the station 'O'Connell's Tower' in honour of Daniel O'Connell.

Thomas Colby, a man of strong will not known for taking advice, had his own ideas on how the mapping project might be conducted. He issued instructions, which became known as the 'blue book'. These instructions were meticulous but proved to be cumbersome and were blamed for the slow progress with the early mapping. His own officers complained about him and he had considerable difficulty with the authorities in England, so much so that a number of enquiries were instigated. He didn't kowtow to his inquisitors but made them witness at first hand what the survey was all about. He took two commissioners of enquiry to visit the triangulation station at Cuilcagh Mountain in Cavan. They had to cross a wide, deep and fissured bog before the arduous climb. They 'toiled, panted and blowed upon their ascent' but were impressed by the professionalism of the work at the summit.

The Ordnance Survey was almost disbanded in 1828 after one enquiry by a parliamentary committee. It was saved from extinction only by a single vote and by the force of Thomas Colby's personality. There were other investigations, one after a complaint from Major Reid, the officer who first administered the work of Ordnance Survey from the Phoenix Park, which eventually reached the Duke of Wellington, now prime minister. Reid's contention was that the mapping survey was hamstrung by Colby's over-zealousness and methods 'so liable to errors of infinite consequence' that additional practices had to be introduced to detect them.

Colby weathered the storms but he had to pay a price. He was ordered to spend at least nine months of every year in Ireland overseeing the work. Up to then he had based himself in the Tower of London and travelled periodically to Ireland. The question of civil assistants and surveyors and cartographers had also been revisited. To Colby's annoyance, Major Reid had begun to recruit them on a small scale in 1826. More were now recruited and by the late 1830s there were three times as many civil assistants as military sappers.

Other preparatory work went on while Thomas Colby and his teams were climbing the mountains and observing angles. Richard Griffith was appointed to head up a new boundary department, the only private surveyor in Ireland to be appointed in an official capacity. His initial job was to delineate the boundaries of all the townlands in the country so that Colby's teams could map them. Griffith took a very pragmatic, some might say ruthless approach to the job, in the way he altered the pattern of territorial divisions. When there were uncertainties or local difficulties, he decided himself where the boundaries lay. He amalgamated smaller townlands into single units and subdivided larger ones into smaller ones.

Townland was the name ascribed to all territorial divisions. The variety of names used for generations faded into memory; 'ploughlands' in Waterford and some southern counties, 'tate' in

northern counties such as Monaghan and 'carvagh' in County
Cavan. Griffith preserved the actual names of each townland
wherever possible, using the names on the landowners' estate maps
or the barony constables' lists. It was Griffith who added the words
north, south, east and west, upper and lower to townlands when he
subdivided them into smaller lots, places like Narrabaun North and
Narrabaun South in County Kilkenny.

The townlands could no longer be mythical, elastic or fuzzy;
there could be no doubt about their position. They had been
staked out on the ground and surveyed on to the map. All future
legal disputes would be settled by reference to the new maps. The
boundaries that had been fluid through the centuries were now
fixed.

A lesser man than Thomas Colby might have been overwhelmed
by the constant criticism he received and the sheer logistics of the
job in hand. Certainly a lesser man would have confined himself
strictly to the dictates of the Spring-Rice recommendations, which
were to produce townland maps of Ireland. Colby adopted a more
visionary and holistic approach to his work.

In 1827 he decided to add geology to his portfolio and instructed
Captain Pringle, who had studied mineralogy in Freiburg, to plan
for the collection of geological information. His plan was halted
during one of the various investigations into the workings of the
Ordnance Survey in 1828 but he recommenced it with enthusiasm
in 1830. Lieutenant Joseph Portlock, noted for his publications in
the new Dublin Geological Society, set up a geological office and by
1839 was employing sixteen assistants.

More significantly, in 1835 Thomas Colby decided to extend the
survey of Ireland to include field boundaries. The six-inch maps of
the northern counties containing only the outline of the townland
boundaries were complete by the time Colby made his decision. The
maps were sparse but functional, delineating the areas of townlands
so that a rateable valuation could be applied to them. Colby did not
have a direct mandate to include field patterns and was in effect
setting aside the recommendations of the Spring-Rice committee.
He took his licence from the Duke of Wellington who some years

previously had told him 'that the maps must be filled up on the scale of six inches to a mile'.

The inclusion of field boundaries made the map more useful, particularly in the event that individual landholdings needed to be valued. This proved to be the case when the 1838 Poor Relief Act provided for the separate valuation of each individual tenement. Through his decision to include field patterns, Colby left Ireland with a priceless picture of the pre-Famine landscape, although the northern counties had to wait until after the Famine for their field patterns to be included. There was high praise for Colby when the six-inch maps began to appear in abundance towards the middle of the 1830s and the dust had settled on the investigations of the late 1820s. Government departments, engineers, even Colby's critics, including Richard Griffith, highly commended the maps. Praise also came from abroad. One German visitor, although not impressed by the level of education of the cartographers in the Phoenix Park, noted that the maps could be ranked among the world's greatest geographical productions.

The first edition of the six-inch map series was completed in 1842 and the final map published in 1846. The final cost was £820,000. Thomas Colby's great work was finished and his system for surveying and map-making firmly established. He had overcome physical and political difficulties on his journey. His health had suffered. He had sacrificed part of his salary to keep costs down, foregoing an annual allowance of £500 for his work in Ireland for the five years prior to the end of his career. His single-mindedness cost him the friendship of Larcom and Drummond. Having been promoted to the rank of Major-General, he retired in 1847 and took his family to Europe to educate his sons. He died 'with scarcely any warning' in England in 1852, at the age of sixty-nine.

Thomas Colby sought neither fame nor honours in his lifetime. He was one of the few of his rank who did not receive a title. His desire, as expressed in a letter to his wife in 1832, was simple. 'When I am no more, I trust those I leave will not think it necessary to pay any pompous respect to the worthless remnant of what once was me. The tribute of sincere affection is all I desire…' When he died, Ireland was the best-mapped country in the world. His successors bore testimony to his work by continuing to update the six-inch maps and give Ireland a variety of maps at different scales up to 1922: the one-inch, the half-inch and the twenty-five-inch.

2

'Intelligent, Lively
but Thoughtless People'

From the beginning, Thomas Colby intended to supplement the six-inch map with additional statistical information on everything related to the country's resources. This information could not be depicted on the map but would contribute to people's understanding of Ireland. He instructed his officers to collect information on communications, manufacture, geology and antiquities. He also asked them to take note of local fairs, markets and the size of the population. This record of social history was called *Ordnance Survey Memoirs*.

Nothing much happened in relation to the *Memoirs* until after 1828, when Thomas Larcom became responsible for the day-to-day operations in the Phoenix Park. An Englishman, he joined the British Ordnance Survey in 1824. But once posted to Ireland he made it his adopted home and quickly demonstrated a great enthusiasm in everything about the country: its history, language, placenames and literature. Larcom spent eighteen years at the Ordnance Survey in Ireland before departing under difficult circumstances at the end of the six-inch mapping programme. Colby had him removed after a prolonged deterioration in their relationship. He remained on to work for the Irish Famine Relief Commission before becoming, like Drummond, Under-Secretary for Ireland.

Larcom enthusiastically extended the scope of Colby's original plans and set out new guidelines for the field officers. As the six-inch maps progressed, they were instructed to collect information on natural features and natural history, modern and ancient topography and the social and productive economy related to every civil parish in the country. With the passage of time, Larcom's plan became even grander. He envisaged extending the work to include the entire British Empire and provide a unique record of the period. But as it turned out he had more than enough challenges in Ireland.

Whatever the merits of extending the memoirs to the rest of the empire, there is no doubt that the work undertaken in Ireland painted a remarkable picture of society before the Famine. The memoir project was carried out between 1830 and 1844. It gives a glimpse of the social and living conditions of the Irish people, their work, habits and livelihoods, their health and education and the mix of Catholics and Protestants. In some parishes local folklore and piseogs were recorded, although not extensively.

Much that was contained in the memoir of each county depended on the attitude and interest of the officers responsible for individual parishes. The consistency, detail and extent of what was written varied considerably. For instance the Enniskillen and Carrickfergus memoirs have very detailed information on all aspects of life and of the countryside whereas others, such as the parish of the town of Monaghan, are quite sketchy.

The first volume of the memoirs was published in 1837. Commonly known as *The Templemore Memoir*, it gave an account of the parish of Templemore in Derry, It contained three hundred and eighty pages and not eight or ten as was first envisaged for each parish. It had cost £1700, more than three times the projected cost for the entire project.

There was widespread support for the memoir project but criticism was initially fuelled by long delays before the publication of the first volume. A draft version of the Templemore memoir did not appear until 1835, almost seven years after Larcom assumed responsibility for the project, and additional information was added before the final publication. It turned out to be a cumbersome volume, difficult to read because the subject matter was poorly organised. There were long historical sections, which were hardly necessary, as this material could be found in other books. Criticism of the content and layout came both from Ireland and England. The Prime Minister, Robert Peel, had no liking for the project.

Criticism also came from surprising sources nearer home. Colby seemed less than impressed with the content, which had been expanded from his original ideas. He commented to his boss, the Master-General of Fortifications, that some of the Ordnance Survey officers were not practised in memoir-writing. This was hardly an endorsement of the project. More pointedly, when he submitted a revised estimate of costs to the Treasury, he gave no recommendation that it should be continued to cover the whole country. Thomas Drummond, now Under-Secretary for Ireland, who had previously criticised Colby's lukewarm support for the project, counselled against the memoir progressing further without the Treasury's positive sanction. His endorsement would surely have carried significant weight with the decision-makers in London. Thomas Spring-Rice, now Chancellor of the Exchequer, who had championed the setting up of the Irish Ordnance Survey, also turned against it. Work on the memoirs was halted in 1840. Further strenuous efforts and a recommendation from the Irish peers failed to convince Peel that the work should be continued. He finally put an end to it in 1844.

This was a pity as only memoirs of the counties of Ulster and small parts of Sligo, Leitrim and Louth had been compiled. Only *The Templemore Memoir* was published. The remainder, all forty volumes, had to wait for a century and a half before being published in the early 1990s.

Had the work continued at the same pace as the six-inch mapping, we would have a total picture of pre-Famine Ireland. We would know how people facing the warmth of the south coast differed from those facing the colder north coast and how those living in the more prosperous east differed from those living in the deprived western regions. We would know how island people differed from mainlanders and how one island's traditions contrasted with those of another.

As it is, Rathlin Island, that most northerly island off the coast of Antrim, is the only one for which an extensive memoir exists.

Were it not for the memoirs, we would not know that of the twelve hundred people who lived there in 1830 more than half had the name McCurdy, and that the Scottish forebears of the McCurdys had fled to the island when Bonnie Prince Charlie was defeated at the Battle of Culloden in 1746. Nor would we be privy to the fact that there were nine hundred Catholics and one hundred and thirty-four Episcopalians living there. We know that two ministers looked after the spiritual needs of the Episcopalians while one priest took care of the spiritual welfare of the Catholics. We learned that the Catholics did not like their previous priest, who, according to the memoir, was Spanish. Evidently they 'held in contempt the unchristian doctrines of their late Spanish priest'. We might have guessed that the poorest on the island were the Catholics but it was a surprise to learn that their houses were made of stone.

James Boyle left a memoir of the island so detailed that he seems to have visited every cave around its treacherous coastline. In Bruce's cave, accessible only from the sea and only in calm weather, he describes how he came across the foundation of a wall above the high water mark. What was its purpose? Boyle leaves it to our imagination. Perhaps it was part of the defences of Robert Bruce, who fled to Rathlin after being overthrown as King of Scotland in 1306. Or was it the hideaway of pirates of the distant past waiting to pounce on passing ships?

I wonder if that was the cave of my childhood history books. It might have been there that Robert Bruce sat mesmerised by a spider spinning its web. He watched, transfixed, as the spider patiently climbed a strand of its web in an attempt to reach the roof of the cave. It failed and slipped down the strand many times but persisted until it succeeded. Robert Bruce took heart from the spider's perseverance and legend has it that he coined the phrase, 'If at first you don't succeed, try, try and try again.' He returned to Scotland to win the battle of Bannockburn in 1316.

Life must have been tough on Rathlin. There was virtually no shelter from the elements, hardly any turf and no trees, although Boyle tells us that there were roots and stumps of oak and fir trees at the bottom of dried-out lakes. The people had to hope that ships from Scotland would come regularly to barter coal for dried kelp.

One wonders if there were many other islands like Rathlin where there was virtually no crime and no illegal distillation of alcohol. The practice of distilling poteen had disappeared many generations before Boyle arrived. The punishment for a misdemeanour might be banishment to 'Ireland' for a short period. The people of Rathlin considered Ireland to be a foreign continent. They had a curse: 'May Ireland be your latter end.' There was no disease on Rathlin. Rheumatism was the only affliction, caused by the harsh living conditions and wet climate. The people had a remedy for their rheumatic pains: their own peculiar sweathouses, small circular stone buildings about four feet in diameter and six feet high in which fires were lit to heat the stones. Up to eight naked men – for this was not an indulgence for women – squeezed tightly into the house, which was then sealed. There they stood in darkness and let the warmth of the stones seep into their bones and drive away their pains. There were two sweathouses on the island, one public and one private.

There was a more comfortable life in the parish of Antrim on the mainland. James Boyle relates how the farmers in general were a prosperous lot who lived in neat stone houses, many with small gardens. Their farms ranged between ten and eighteen acres in size. Their diet was good: bacon, salt and dried meat, Lough Neagh fish, baker's bread, eggs, milk and potatoes. They had abandoned baking oaten bread at home as baker's bread was cheaper. Their methods of farming were advanced. Boyle's only negative comment was: 'they are rather addicted to whiskey-drinking but are otherwise frugal.'

Other memoir-writers did not describe such order and harmony. The memoirs of Monaghan and other counties tell an entirely different story. There the houses and estates of the landlords contrasted sharply with those of the tenant farmers who leased smallholdings that were scarcely big enough to sustain a family.

Lieutenant Taylor tells us that the five hundred acres of land owned by Lord Cremore in the parish of Ematris 'ranks among the most handsome nobleman's seat within the province, adorned with plantations and ornamental grounds of great extent and beauty.'

This contrasts with his description of the houses of the tenants: '...the mud houses divided into three apartments seldom exceed one storey high, furnished occasionally with glass windows but often without them, an earthen floor with no ceiling, and universally thatched with straw, one extremity apportioned a bedroom for the family, the opposite for the cattle, and the centre a kitchen and dining room for the whole household.'

'Comfort and cleanliness are little observed by the peasantry of Ireland. In truth, nothing can surpass the filth and dirtiness of their cabins and the enclosures around them.'

It was rare to find three rooms in the poorer houses. Most had only one, or two at the most. The memoir of the parish of Dungiven relates how the population of Benada Glen was very dense and that it was common for three generations of the same family to sleep on a single bed of straw on the floor of their one-roomed cabin. In one house there must have been a conventional bed because the writer notes: 'the bed is for the granny'.

The memoirs of Monaghan go on to reveal many of the causes of this state of affairs. The landholdings, between two and ten acres, were not large enough to support the average family of two adults and five children. There was little incentive to improve those holdings, which were held on short leases or at the will of the landlord. It was likely that rents would be increased if improvements were made to the land and tenants evicted if they could not afford the increased rents. As it was the rents were invariably high and any cattle or pigs reared or oats grown went to pay them. This left nothing but potatoes as the staple diet of the poor.

Lieutenants Taylor and Chaytor, the two principal writers of the Monaghan memoirs, blame absentee landlords for most of the ills of the Irish poor. They leave us in no doubt that the condition of the tenants could not be improved while landlords continued to let land on short-term leases and until the practice of subdividing farms into smallholdings ceased.

However, Taylor also testifies to the resilience of the poor when he writes about the sale of all farm produce to pay rents demanded by the landlords: 'All are transported to make up the rent and nothing remains but the light, gay and cheerful spirits of the emaciated frames of the half-starved population.' He tells us 'that the healthiness of the people is well established by the medical practitioners' and that longevity is common. Virtually all the memoir-writers tell of many people in every parish living to one hundred years or more. It is difficult to believe that they could live so long in such poverty, on a diet of potatoes and buttermilk and in squalid dwellings. It had be their cheerful spirits that kept them alive.

In some parishes, like Kilmactigue, these 'intelligent, lively but thoughtless' people had dealings with another sort of spirit. Lance-Corporal Trimble believed that all the farmers of that parish used their entire grain crops to make 'poteen whiskey' and 'drank as much of that liquor as would defray the expense in sending their oats to Sligo properly for sale.' Trimble gathered a substantial knowledge about the parish for he could record that 'there is upward of 400 stills at work'. He also knew that the poteen and grain from which it was made were hidden under turf clumps in bogs, under the bottom stones of lime kilns, in potato pits and under the hearthstones of kitchen fires.

John Stokes discovered the ingenuity that was really under the hearthstone when he was writing about the countryside around Magherafelt and Dungiven in County Derry. There was often a hidden room excavated underneath the hearthstone, where poteen was distilled. A small hole from outside allowed a stream of clean water, a critical element in the distilling process, to flow into the room. This hole also let in light and air. The smoke from the still was funnelled up into the fireplace of the house so that it mixed with its smoke. In that way it became virtually impossible to detect the presence of the still. Stokes also discovered that wandering distillers travelled from place to place plying their trade but never resting for more than a day lest they be apprehended by the police.

Despite this illegal operation, Taylor reports that crime had decreased in 1836, although Chaytor tells us that 'card-playing and a strong party feeling existing, causes much quarrelling and disturbances' in the town of Ballybay, County Monaghan.

For all the oppression, poverty, hardship and merrymaking, the memoir writers tell us that the Irish were anxious for learning and attended formal schools. Many of the schools were funded partly by student contributions, which amounted to between sixpence and one shilling and sixpence a quarter, and some children attended hedge schools. They disagreed about whether or not the discipline of the schools had helped to improve the morals of the people. It appears that most schools did contribute to the decline of the Irish language at that time. We are told about a school in the townland of Termeel in County Derry which had seventy children and in which the teacher was astounded at how the pupils learned English perfectly in a 'couple of months'. This particular school did not last long as it was denounced by the priest because the teacher was Protestant. Stokes records that priests also interfered with the running of the schools by objecting to the teaching of Irish.

The Catholic clergy had an active influence on other aspects of the changing face of Ireland. They attempted to stamp out many of the superstitions that had come from ancient pagan traditions, including those associated with holy wells, but this did not deter people from secretly placing votive rags on bushes and branches as petitions to heal ailments, as was the custom at Tobar Phadruic near the graveyard in Dungiven. Nor could the clergy prevent people from believing that the compasses used by the sappers had supernatural powers. Some believed that the compasses were activated by fairies, while people in the parish of Upper Cumber believed that a sapper carrying a compass could foretell future events.

In Dungiven, the ancient 'keeny and taisch' at wakes and funerals were also in decline and being replaced by hymns such as the 'Day of Wrath', hardly a fitting replacement for the ancient Irish rite of lament. John Stokes tells us that formerly the keeny or *caoineadh*, a high-pitched scream of women mourners, was sung at wakes. The taisch or *tásg*, meaning a report of death, where singers responded to each other in high-pitched screams, was sung along the road to the funeral. John Stokes laments its disappearance and suggests that Catholic clergymen should 'encourage and revive a little of the taisch'. He believed that 'when sung alternately with the hymns, the plaintive modulations' of the taisch 'would tend to produce the sadness in the minds of the hearers most in keeping with the ideas' of the hymns. The custom had not died out fully for it was still practised in remote and mountainous parts of the country but presumably it was only a matter of time before the new fashions would reach these places.

3

THE WITCH'S NOSE AND OTHER PLACENAMES

Thomas Colby was faced with the challenging task of determining the official anglicised forms of the placenames in Ireland. Each of the sixty thousand townland names, as well as other significant names, had to have an anglicised version set down on the first six-inch maps of the country.

The townland names had, for the most part, their origins in Ireland's distant past. Spellings and pronunciations had changed over time and in many cases had mutated to English words that were barely related to their origins. Who, for instance, would have guessed that the townland of Strancally in County Waterford derived from *Srón Caillighe*, which means The Witch's Nose? Or that Kilwatermoy, also in County Waterford, had its origins in *Cill Uachtar Maighe* ('The Church of the Upper Plain')? Other townland names had disappeared from use altogether, perhaps because the English planters believed them to be 'barbarous and uncouth' (the words of King Charles I), and townlands were given the landlord's name. The townland of *Cill Cronagh* in County Kilkenny became Greenville because it was the home of the Green family and *Baile an Fhásaigh* (The Town of the Wilderness) in the same county, the seat of the O'Shea family, was renamed Sheastown.

It must have been extremely difficult for the Ordnance Survey's engineer officers in the late 1820s to deal with Irish placenames. None of them had previous knowledge of Ireland and certainly not of its language. Even if they had, accents and dialects would have confused them. They were also faced with an Irish language that was in decline and which, in some places, like parts of County Down, had disappeared completely. In the northern counties they had the Scottish influence, brought to the names by immigrant settlers, to contend with.

The officers set out the many variations of townland and other names in the *Name Books*, giving the spelling and the authorities from which they came. Their sources were diverse: official records like the barony constabulary lists, the landlords' estate maps and lists compiled by Richard Griffith's new boundary department and, of course, existing maps. Clergy and schoolmasters also contributed information. But the officers did not consult with the native people who were closest to the names.

Colby took a pragmatic if unscientific approach to deciding which of the various spellings of a name he would select. He chose the most commonly occurring spelling from the *Name Books*.

Larcom took a different view when he assumed responsibility for placenames in 1828. He had a greater interest in the Irish language and in placenames and wanted the anglicised names on the six-inch maps and those included in the memoirs to correspond as closely as possible to the original Irish form. He wanted to take a hands-on approach to the subject and decided to learn the Irish language. To help him, he engaged an Irish scholar, John O'Donovan, three times a week to teach him Irish, hoping that his knowledge of the language would develop sufficiently to enable him to make decisions on the proper spellings of placenames. He quickly discovered that this idea was impractical and that he really needed an Irish linguistic scholar on his staff, so in 1830 he employed O'Donovan as a full-time member of the Ordnance Survey. O'Donovan's name would become synonymous with Irish placenames.

O'Donovan was born in 1806 in Atateemore in south Kilkenny, which in Irish is *Áit an Tí Móir* (The Place of the Big House), and went to school in Patrick Street in Waterford, where he learned arithmetic, English grammar and bookkeeping. This was not the most salubrious part of Waterford: in the *Name Books* O'Donovan describes Murphy's lane off Patrick Street as: 'A paved lane. Its occupants are of the lowest class of tippers, whores and pick-pockets of the lowest most diabolical character, so at least they were in 1821 when I went to school to old Ned Hunt of Patrick Street.' He subsequently learned Latin from a monk, before moving to Dublin at the age of seventeen: there he studied Latin, Greek, French and Italian at Saint Patrick's Seminary on Arran Quay. He became an expert in the Irish language and in Latin.

O'Donovan brought a new dimension to establishing the anglicised versions of placenames for both the memoirs and the maps. With a team employed in a new topographic department, he studied the ancient sources of placenames in manuscripts and books in libraries and universities. The names from these sources were added to those collected in the *Name Books*. O'Donovan travelled extensively throughout the country, starting in County Down in 1834 and ending in County Kerry in 1842. There were three counties he did not manage to visit – Antrim, Tyrone and Cork – although he did work on their placenames. He met elderly Irish-speaking people and listened to their pronunciations. He also studied the names in the context of their association with the old topography and antiquities of the area to give him further clues about their origins.

He settled on spellings that were nearest to the pronunciations used in the district. For instance, he acknowledged that the word *cluain*, meaning meadow, which forms part of many placenames, had different pronunciations in various parts of the country. Hence we have Clonmel in County Tipperary, Cloonty near Baronscourt in County Tyrone and Cloyne in County Cork, all originating in the same word. The spellings O'Donovan derived did not always correspond to the older written sources or those recorded by the officers in the *Name Books*. John Andrews tells us in *Paper Landscape* that in a random sample of one hundred O'Donovan spellings, forty-six differed from all recorded authorities. But the names O'Donovan derived became the definitive names of townlands and were solidly engraved on the maps. They would form the legal definitions of places for future generations.

O'Donovan had a significant advantage over the engineer officers who collected the first tranche of placenames in the *Name*

Books. He was Irish and recognised as a learned scholar. He sought and received the patronage of both clergy and landowners and in many instances gained access to private libraries and papers. The Bishop of Down, Dr Crolly, supplied him with a letter of introduction to the priests of the dioceses in which he urged his clergy to give 'every assistance which they can afford, in order that Mr O'Donovan may be enabled to accomplish his interesting and important object'. But his first attempt to meet the bishop's clergy, who happened to be at a gathering in Newry, was not successful. They were obviously enjoying themselves thoroughly, for he thought it would have been 'impolite to interrupt such a jovial and convivial class of his Majesty's subjects with his dry specialisations'. He discovered later that they could have been of no assistance to him. He did not always get the cooperation he desired. Some people were reluctant to give him information in case he was a government spy or because they might be called up for military service. In County Cavan he was refused accommodation in a number of lodgings because he wore black clothes and looked like a Methodist clergyman.

County Down was not an easy county to begin with. Even though he was a great Irish scholar, O'Donovan experienced many frustrations in determining the origins of some placenames. His ear, attuned to the accents of the people of Dublin and counties further south, was not accustomed to the northern accents. In March 1834, he wrote from County Down: 'I have today travelled through the parishes of Hollywood and Dundonnell and found the inhabitants, who are Presbyterians, knowing nothing of names…Their pronunciations are so barbarous that it is very difficult to catch the sound. I find that it is absolutely necessary to get a gentleman (alias a country rich savage) and a plebeian to pronounce them. The sound is not English nor Irish nor Scotch but a chaos of the three and in which the Scottish accent predominates. Every Bally is pronounced Belly and it is almost impossible to know from their pronunciation whether the vowel should be a or e.' In the same year in Lisburn an old priest told him, 'I am afraid you are a hundred years too late.'

John O'Donovan and a small group of assistants, among whom

were the Irish scholars Thomas O'Connor and Eugene O'Curry, collected far more than placenames in their travels. They recorded detailed descriptions of antiquities: old parish churches from which the name of a townland or parish might have come, old castles, cromlechs, raths and forts. The descriptions included both the measurements of the buildings and comprehensive accounts of what remained. William Wakeman, who had been employed to work on the hill-shading for the proposed one-inch maps, followed O'Donovan and made drawings of the more important antiquities.

We know from their records that the round tower in Ardmore, County Waterford, originally had a 'cross like a crutch' on the top of the cap of the tower but that it was gone when they visited. Some years previously it had been taken down by 'repeated discharges of musket balls'. And we know that the door in the ruined church of Mothal in the parish of Kilmademoge, County Kilkenny, was closed up to turn the church into a handball alley.

O'Donovan and his companions describe the traditions associated with many of the old parish churches and the date on which 'patterns' and 'stations' took place. Many of these old customs were gone, having died out or been forbidden by the clergy over the previous thirty years. The parishes of Aglish, Kinsalebeg, Croake, Kilmacombe and Kill St Lawrence in County Waterford had abolished the customs. But we know from O'Donovan's letters that patterns were still held in some places. In the parish of Rathgormuck the pattern was held on 29 September each year in honour of the patron of the church, St Michael the Archangel. In O'Donovan's description of Rathgormuck's old, ruined parish church he tells us that the 'west gable is featureless but is covered with rags inserted into the crevices by people who perform the stations in and around the church'. At that time, about thirty people participated in the stations there.

The recording of the antiquities was not strictly part of O'Donovan's work but he was encouraged to do it by Thomas Larcom. The work was timely as many of the antiquities were being destroyed or vandalised and were fast disappearing. A schoolmaster in County Down, Robert McVeagh, pointed out the site of 'one of the finest forts in this parish' that had been levelled some years before. He told O'Donovan that he was aware of twenty-four forts that had been levelled in the parish. In these cases they were levelled for a very practical reason. The tenant farmers needed to cultivate every square inch of their land to pay the rents and could not afford the luxuries of unproductive ancient monuments on their land.

It is extraordinary to think that O'Donovan carried out such extensive work on the placenames of the entire country in such a short time and that he walked the hills and valleys in all sorts of weather. Transport was not easily available throughout the country and what there was left a lot to be desired. O'Donovan recounts many miserable journeys. On one of them, in April 1835, he travelled on a coach from Derry to Enniskillen, a distance of some sixty miles. It was an eventful journey, sitting in the open with 'seventeen noisy, inebriated fellows'. Umbrellas were shattered in the gusting wind and he was soaked to the skin. In Strabane, searching for a glass of whiskey to revitalise his circulation, he lost his seat to new unruly travellers and had to sit on top of a trunk. Further on, the coach overturned because of a careless driver and he was thrown into a ditch. Luckily, he escaped with a few minor injuries. He had to find lodgings for the night. Next morning his seat had been taken and the coach was overloaded so he walked the final twenty-two miles to Enniskillen.

Writing from Dromore on Good Friday 1834, he said, 'I find travelling on coaches or cars will not accelerate my progress, and therefore I prefer walking because it will be less expensive to the public and myself.' The reference to himself may have been meant as a swipe at his paymasters. He constantly complained about his poor pay throughout his years working for Ordnance Survey. His starting pay was two shillings and sixpence a day. He was compensated for his travelling but not his lodging expenses. In fact he resigned his job briefly in 1832, partly because of his pay. He was not paid during periods when he was unable to work because of sickness, even though his illness was the direct result of the many hardships he endured. Neither did his paymasters take into consideration the long hours he worked.

He spent December 1838 and January 1839 travelling throughout Wicklow with Thomas O'Connor. Again the issue of money
arose when he decided in the dead of winter to walk from
Blessington to Glendalough because 'the Hotel Keeper would not
send a car thither at the usual price and I was not willing to give him
more.'

It was a foolish decision, for when he and O'Connor attempted
to cross the mountain they ran into a snow blizzard with nine
miles still to walk. They persevered until a 'countryman' told
them that most of the road from there on was uninhabited and
that two bridges had been washed away in a previous storm. He
called himself a coward for turning back and walking three miles
to 'Charley Clarke's public house, where we got infernally bad
treatment'. The bed was damp and the room 'horribly cold'.

Next morning they attempted the journey, despite O'Donovan
having developed a bad fever. They climbed through the soft
snow, 'regularly sinking thro' the half dissolved masses of snow
and occasionally down to the knees in ruts in the road. One of my
shoes gave way and I was afraid that I should be obliged to walk
barefooted.' They made it to Glendalough that day. He describes the
scene as 'horribly beautiful! and truly romantic, but not sublime!'
The fever and hardship of the day did not deter him. He purchased
'a pair of woollen stockings and knee breeches, and went at once
to look at the churches, which gave me a deal of satisfaction'. That
night, more misery lay in store for him after 'a very bad dinner.' It
was the night of the Big Wind of 1839 and the windows of his room
were blown in on top of him. A violent gust tossed him across the
room while he was attempting to keep the shutters closed with the
weight of his body.

Unfortunately, O'Donovan had to truncate his account of the
storm because he ran out of writing paper. 'Pity I have not paper
to tell the rest,' he wrote. This was a regular occurrence, which
prevented him from compiling reports. Supplies were regularly sent
to him from Mountjoy House but very often did not arrive at their
destination until he had moved on, or they were stolen in transit.
To overcome some of the difficulties he arranged that two boxes of

stationery, ink and quills be dispatched each time and sent to his anticipated destinations.

So now as I drive or walk through the parishes and townlands of County Waterford, I imagine John O'Donovan and some of his fellow travellers like Eugene O'Curry or his brother Anthony walking the roads, seeking out the ruined parish churches and cromlechs and raths. I wonder about the location of the dark, damp, cold lodgings where they spent hours writing by candlelight at night. When I see a thatched house, like that of my cousins in the townland of Smoor, I wonder if a *seanchaí* had previously lived there and if he had relayed to O'Donovan the stories and folklore of the parish and pronounced the placenames as handed down to him. I can hear John O'Donovan repeating the placenames and contrasting them with what was written in the *Name Books*. I wonder what his comments were when he noted that Smoor, *Smúr* in Irish, means rubbish or dross.

I think of him as I approach a bland roundabout called Mona-mintra on the Waterford City ring road and ponder the significance of the Irish version of the name, *Móin na mBaintreach* or the Bog of the Widows. I look around to see where the bereaved women might have toiled and find no trace of what might have been a bog. When I drive through the parish of Kilburne I recall that John O'Donovan reminded us that in 1841 the natives called this parish *Teampall a' Chnuicín* (The Church of the Small Hill). And I ponder his description of the cromlech in the townland of Knockeen in that same parish, something I marvelled at when I was a boy. I am proud that he too visited it and wrote that it is 'one of the most remarkable monuments of pagan antiquity in the county, if not in all Ireland'.

When I leave the Tramore road and travel through Towergare, *Tuar Gearr* (Short Bleach), I recall that, according to the 1842 six-inch map, there was a rath or fairy fort here. Like the forts of County Down, it disappeared to give way to farming; only now it is modern farming. And I cannot help remembering John O'Donovan when I walk on the beach at Garrarus or *Garbhros* (Sharp Promontory), along the County Waterford coast, and reflect on how accurate the description is.

When I stroll into County Kilkenny from Waterford, across the bridge over the River Suir, I pass through the townlands of Mount Misery, Mount Sion and Christendom. For as long as I can remember, I was fascinated by the strange names and wondered about their origins. They sounded so Biblical, I was sure there had to be a religious connection. But I was disappointed when I discovered that the townlands were called after the names of the big houses situated in them. O'Donovan was not impressed either, for he caustically remarked in the *Name Books* that these were 'fancy names'. However, as I walk a little further on, I am happy to discover that the townland of Belmount also had an Irish name. O'Donovan records that the ancient people of the area knew it as 'Ballynagraige or *Baile na Gráige* (The Town of the Crags). He recommended that the name would appear on the six-map along with Belmount but it never did. His home townland of Atateemore is nearby. It is also called Blackneys, after the big house, but somehow he ensured that its Irish name also appeared on the map.

Again, when I travel in west County Waterford through the Barony of Glenahiry or Glennahira, as O'Donovan spelled it, I pause to see what he meant when he wrote that 'This is the most Irish county I have yet traversed and I am sorry to say the less interesting in its antiquarian remains; and the people are not as enlightened as any county in Connaught.' And I am happy that the people of this area, like the people of the rest of Ireland, have emerged from the poverty of the nineteenth century and bear no resemblance to O'Donovan's description of those who lived there then; 'The land is craggy; poor; unproductive; and the people are overworked; they are actually stunted or twisted (*casta*) in the earth with hard labour and still they are not rich nor happy, but very quiet. Their martial ardour is dampened!'

Recently, to my amazement, I discovered that John O'Donovan's forebears belonged to the parish of Kilmacow in County Kilkenny, where I now live. His ancestor, John Donovan, lived here in the early 1700s and his grand-uncle, Edmond Donovan, was its parish priest. Five generations of his family since 1735, including his father, are buried in the churchyard in Dunkitt, a mile from my home.

4

THE RELUCTANT RECRUIT

It was 1970 and I had just graduated as a civil engineer when I was offered a job with the Ordnance Survey and a promise of an exciting future with loads of travel both in Ireland and abroad. I was told that I could become director some day. There would be adventure. I could travel and explore every road and boreen, every field and mountain, every historic site and modern town in Ireland to update and redraw the maps of an entire country.

At first I did not accept. Jobs for civil engineers were plentiful; Ireland's urban landscape and communications infrastructure were rapidly developing. There were more attractive jobs on offer from local authorities and exciting prospects with consulting engineers and construction companies. The Electricity Supply Board offered me a one-year scholarship to return to college to study computers and soil mechanics and map the behaviour of soils under stress. I accepted their offer. During that time I was sucked into a fascination with mainframe computers. I spent endless nights, staying up until dawn, programming in the college computer room. I could have moved on from there to a more challenging job in the developing world of computers.

Gerry Madden, the Assistant Director of Ordnance Survey, sought me out again fifteen months later and said he had kept the job open as he thought I was interested. This time I accepted, having finished my studies and spent two boring months working in an overheated ESB office in the centre of Dublin and having married his niece in the meantime.

I did not have any great knowledge of the job other than what Gerry Madden had told me and I seemed to drift into it without any motivation. I knew that the Ordnance Survey offered a good pension, a job for life and prospects of promotion. Perhaps it was my upbringing and the conservative streak in my mother, who always advocated security for her children, that influenced me. Or it might have been some innate insecurity I carried from childhood. My mother had bought some of the groceries on credit each week and paid off part of the debt with my father's work bonus at the end of each year. The grocer gave her a year-end discount which wiped the rest of the slate clean. I had a childhood dread that she might not be able to clear the debt.

Or maybe something deeper was inclining me towards the Ordnance Survey – that childhood fascination with the mono-chrome map of Waterford City and the one-inch coloured map of the county.

I had to join the army, for that was the only method of entry to Ordnance Survey. I would have to become an officer in the Survey Company of the Corps of Engineers. I never had a liking for the army, never had an itch to join it and here I was attending for interview at a military barracks. It was November 1971 and the army was on high alert because of the increasing violence in Northern Ireland as the nationalist population demanded civil rights. There had been talk in the newspapers about the Irish army invading the North to protect the nationalists. I didn't fancy the idea of going to war.

I sat in a bright office with the sun beaming on my face through a high window. Four stern-looking army officers sat across a table from me in full uniforms, brass emblems on shoulders, medals on chests and shining Sam Brown leather belts. I didn't know who they were other than that one was the Director of the Corps of Engineers. As for rank, I didn't have a clue. I was aware of titles – Colonel, Commandant, Captain, Lieutenant – but I was ignorant as to what a person of any rank looked like or what insignia he might wear. The only images that I had in my mind were that a colonel would be old, overweight and probably full of pomp and that a more junior officer might look a bit smarter and his uniform might be a better fit for him.

The interview got off to a bad start. 'Have you any relatives in the army?' the grey-haired officer in the centre enquired.

I was affronted by this question as I took it to mean that I would be more favoured if I had. I had a flashback to something my grandfather often repeated when I was very young: 'Earn what you get on your own merit.' He had lived by that premise and I did not want this job if I was to be judged only on who I knew.

'I have none,' I said, even though I was related to General Richard Mulcahy, a former chief of staff and Minister for Defence, and was now, through marriage, related to Gerry Madden – and, I found out later, to the man who asked the question. He was not aware of this.

'Why do you want to join the army?' the same officer asked.

'To work in Ordnance Survey,' I replied. I had no other motive, no patriotic or higher calling; I looked on it simply as a pragmatic way of achieving a goal.

'What would you do if you were not guaranteed to be sent to Ordnance Survey?' the officer continued.

'Then I would not join the army,' came my swift reply. I might have been unsure about why I wanted a job in the Ordnance Survey but I was certain that I did not want to be an army officer for the rest of my life.

There was silence. The officers at the far side of the table looked at one another in disbelief. The officer who had been busily writing also stopped. I did not see anything wrong in what I had said; it was the truth. The office who had spoken already made another effort at a question although, this time, it was perhaps more a statement: 'You might not have a choice in where you are sent in the army.'

'Then I won't join the army.'

There were no further questions. No one asked if I had any questions. I was told they would be in touch with me in due course.

Evidently my interview did cause a stir. Years later I was told by a friend that the officers who had interviewed me had been furious that a young whippersnapper graduate should show such arrogance towards senior military men, such impertinence to army colonels!

This was 1971 and I had barely left my student days, when freedom and self-expression were the order of the day. Like most other students I had carried the *Little Red Book* of the sayings of Chairman Mao Zedong and flirted with communism. Young graduates were carefree. Wisdom was not then part of my makeup. I was too inexperienced to know that games and politics had to be played and that sometimes what is said or asked is not what is meant. I had not yet learned the language of coded messages.

I forgot about the interview and continued to work as a design engineer with the Electricity Supply Board but I was bored and began actively to seek other employment. Six weeks after the interview I received an unexpected telephone call. A young lady telephoned me saying she was from the Department of Defence and reminded me that I had sat an interview for the Corps of Engineers. 'There is one problem,' she said. 'It has gone on record that you stated that you would not join the army unless you were sent to Ordnance Survey.'

I informed her that I did not see anything wrong with that and that I was only interested in joining the Ordnance Survey. In vain, she tried for ten minutes to convince me that I could not dictate to the army authorities where I wanted to serve. Eventually, probably at the end of her tether, she sweetly said, 'Look, if you say you will go where you are directed, then I guarantee you will be sent to Ordnance Survey.'

I thought it rather strange that I had to go through this charade to get a job. I questioned her about it and she reassured me that was the case – and I agreed. Two months later I knew what a First Lieutenant was – I was one, having spent a cold November and December in the misery of the Curragh with the Corps of Engineers:

Cropped grass, furze bushes, sheep and sheep droppings, tufts of dirty wool everywhere.

Stacks of turf, smell of smouldering open turf fires inside and outside.

Soldiers, officers, saluting, square-bashing, regimental order, obedience.

Old, neglected buildings, splendid officers' messes.

Christmas was worse. It was my first married Christmas but instead of being with Eilís, my new wife, who was pregnant, I was stuck on guard duties in the greyness and dampness of a barracks. I had the dubious privilege of being the most junior officer in the Corps of Engineers and this was my reward. Later I found out that I had been posted to the Ordnance Survey as technical officer in the survey company before Christmas and should not have been near the Curragh, but no one had informed me.

I was glad to get out of there. I often wondered afterwards why I had not run from that desolate, cold place.

5

A Place that Time Had Forgotten

I was full of anticipation when I finally completed my training in the Curragh and passed through the heavy green gates of the Ordnance Survey in January 1972. It was a place that time had forgotten. Bare grey branches of beech, oak and chestnut trees almost hid it from view. A surrounding moat protected it from the advances of the changing world. There was decaying grandeur all about. Mountjoy House, built in the eighteenth century, its front clad in purple ivy, red-brick offices fashioned from an early-nineteenth-century cavalry barracks, and cut limestone buildings from the late nineteenth century surrounded a shabby cobbled courtyard. There was little sign of life apart from a spattering of haphazardly parked cars on the cobblestones.

Gerry Madden was waiting for me in his office in Mountjoy House. It had once been the home of Lord Mountjoy. Older staff swore to me later that the house was haunted by his son, who was killed in the Rebellion of 1798. Senior men who worked late at night were adamant they had seen him many times, silently climbing the basement stairs. Some were too frightened to return there at night. It was easy to believe. Doors drooped on their hinges. High ceilings in musty halls were complete with plaster cracks and flaking paint. Intelligent spiders made webs in places brooms could not reach. Patches of dust rested on the uneven plaster of cream walls, which were long faded from age. Floorboards creaked and stairs had long ago shifted from original positions to rest in independent ways.

White plaster busts of the founding fathers of map-making

in Ireland – Thomas Colby, Thomas Drummond and Thomas Larcom – stood on sombre black plinths in the hallway. They had presided over sixty years of decline. Their ghostly faces looked on despairingly as dank musty air rose from the basement stairs and floated past their nostrils. The records of the work that had been so dear to them had been stored to gather mould in the damp and blackness of the basement. Rats, mice and cunning cats were common visitors to that place.

By the time I entered Gerry Madden's office I was already wondering what madness had brought me to a crumbling, stale building built in 1728 and why I had not carried out a little more research before joining the army. I hoped I would find better in other parts of the Ordnance Survey. Gerry Madden did not seem to notice the surroundings or that a new recruit might not be impressed. He was enthusiastic about the future. There was to be a huge increase in staff. There would be eight hundred and fifty people working in Ordnance Survey within a few years. There would be new maps of all the cities and towns within ten years. The rural maps would take longer to complete. I asked how long. I had to ask twice before I received an answer – twenty-five years.

He was not inclined to dwell on the recent history of the Ordnance Survey, although he did tell me that the maps had been left to deteriorate after 1922 because nobody outside the Ordnance Survey saw the need to do anything about them. Politicians and administrators assumed that the landscape would change little. A cynic might have said that there were no votes in maps. In any event, the economic state of the country in the first half of the twentieth century was such that there was no money to spare for the luxury of updating them. Only a few more than a hundred people worked in the Ordnance Survey after Irish independence, compared to three hundred immediately before – and they did the best they could.

I really did not have to ask him anything: the buildings and a tour of the map-making process told me everything. Later, someone told me that money had been so scarce in the 1940s and early 1950s that pencils were often divided in two. Sometimes cards were played to decide who would do the day's work: the winner always got the job. I suspected that Gerry had known this and that it had happened in his time too.

We climbed a creaking staircase and walked along narrow winding corridors on uneven floors. Our first port of call was a dull copper plate showing a map of part of old Dublin which hung on one of the corridor walls. I could see Gerry had told the copper plate story many times. He told me that all the original maps had been engraved in copper. 'But look,' he pointed to the names. 'They were engraved in reverse, as was the entire map.' The maps had to be drawn in reverse so that the images printed from them would be right-side up. It was tedious work and the cartographers engraved only one square inch a day.

There was folklore attached to the engraver-cartographers. They were the elite. They arrived at work each day in horse-drawn carriages. They wore bowler hats and grandfather watches in their waistcoats and carried hip flasks filled with whiskey to fortify them for the day. I could imagine them arriving into the courtyard on that very day.

When we entered the drawing office Gerry had something new to show me. The drawing office was like the rest of the building: antiquated and in need of painting. As I stepped inside, I could not help but feel I was walking into some medieval monastery where silent monks toiled endlessly in a scriptorium. Men sitting on high stools – for there were only men there – bent over tall desks meticulously cutting the outline of fields and houses into plastic sheets coated in some soft white material. The lack of a female presence and touch in the offices was striking. There was neither tidiness nor decorative taste anywhere to be seen; flowers and ornaments were absent. Instead battered kettles and stained mugs kept one another company in corners of rooms; lunch boxes and spoons and bags of tea and sugar adorned window sills.

Women could not work here. There was no way for them to be employed, as joining the army had been a prerequisite to being employed in the Ordnance Survey since the 1930s and as yet the army could not recruit women. Miss Grace, a typist, was the only exception. She plied her trade on a mechanical typewriter in the administrative office. There was no ladies' toilet in the building and she had to walk across the cobblestone square in rain and sunshine to find privacy in the resident superintendent's house.

I was astounded to see not an image of a map on the wall of the Ordnance Survey's drawing office but a deathly-white impression of Tutankhamen, the Great Pharaoh of Egypt. It was a training piece for the new plastic which was replacing paper as a drawing material. The cartographers were busily drawing new maps of Dublin on the plastic sheets and Gerry was very proud of it. I was not impressed. The maps and the image of the pharaoh lacked life. The cold precise lines cut into the plastic were a far cry from the soft lines engravers had crafted a hundred years previously into the copper plate I had just seen. Nor did they bear any resemblance to the uneven lines a draftsman's ink might have imprinted on paper a few years before.

There was one thing which did impress me as I passed from office to office: the pride of the cartographers in their craft. I could feel their emotion as they explained the intricate details of what they did. Generations – fathers, sons, nephews and cousins and friends of friends – worked here, recruited by word of mouth, hand-selected, knowing that tradition would be preserved by the next of kin, sure that the skills would run in the veins. They preserved the process of map-making as it had been in Thomas Colby's time, although small variations and tinkerings, new materials and improvements in equipment had made some things easier.

One beam of light, giving hope for a brighter future, shone through the haze of past glories and ingrained traditions I had seen throughout the offices. Aerial photography, which had been introduced by the Ordnance Survey in 1965, was being used to make new maps of Cork and Dublin. One very young cartographer in military uniform peered into binoculars on a cumbersome army-green machine and turned hand wheels. Another equally young man ensured that a pencil on an attached drawing machine remained sharp enough to draw houses, roads and rivers.

Gerry Madden assured me that this was the future for Ordnance Survey. He repeated what he had said in his office, that all the towns and cities of Ireland would be newly mapped this way within ten years. I could not imagine how a single machine could make inroads into so many places.

I wondered how the senior men, steeped in the traditions of the place, had taken to this new technology. John Danaher, the man responsible for air survey, told me later that there had been many critical murmurings when mapping from aerial photography started. Older men, immersed in the methods of the past, reacted in disbelief. They could not credit that an operator could look through a pair of binoculars into a cumbersome machine and see a three-dimensional view of the land. Less still could they comprehend how he could draw a map of the landscape. Many of the men shook their heads and mumbled, 'Only the eye of a man in his wellington boots can interpret the land. The machine can only draw lines and it could never be as accurate as the trained eye.'

After a few months of familiarisation at Ordnance Survey I departed for another army, the British Army, to undertake a postgraduate course in mapping and surveying. The internationally recognised course at the School of Military Survey in Hermitage near Newbury lasted for eighteen months. It seemed strange to be sent to an English military establishment in 1972, when relations were tense between the Irish and British governments. The Northern Troubles and civil rights movement were at their peak and the events of Bloody Sunday in Derry that January increased tensions further. Gerry Madden briefed me about keeping a cool head if Anglo-Irish issues were debated.

But there was no animosity; it was all very polite. The question of Northern Ireland was never discussed. Anglo-Irish relations were never mentioned. It was as if this was another world, dissociated from politics or any messy business, an independent world of pure map-makers, experts who let the facts of the landscape speak for themselves. Colby's tradition – that sappers would not become involved in the troubles or politics of the country they were surveying – lived on.

The school was part of the Royal Engineers. Both military and civilian instructors ran the course. The group on my course included English, Canadian and Thai army officers, along with civilians from Uganda and Hong Kong. We were all, military and civilian, addressed as Mister. Only the military instructors wore uniforms. It was less formal than I expected for a military establishment, perhaps because of the mixed status of the participants.

There was one thing I could not fathom. The military hierarchy extended to the wives of officers. The Colonel's wife was, like her husband, a commanding officer, and the rest of the wives were regarded as having equal rank with their husbands. She dictated the social lives of the women. Attendance at coffee mornings and other recreational events, like strawberry picking and visiting castles, was virtually mandatory for all wives.

It was an engaging and intensive course, a far cry from my short surveying course when I was training to be an engineer. Every hour of the day was occupied in true military fashion. The theory and practice of mapping and surveying were taught from first principles and with antiquated machinery and instruments; surveying computations with cranking hand-held calculators, field surveying with chains, drawing maps with pre-Second-World-War air-survey machines.

We even studied astronomy. I remember very well spending hours on either side of midnight of a freezing January night in 1973 observing the stars to determine our latitude and longitude. I could not wear gloves because I had constantly to twist the fine-motion screws on the theodolite to track the movement of the stars. Everything here was from a past era. The equipment was old; the methods of map-making were old. There was little to give me hope that map-making might move into a modern era. But I persevered. I went to England alone and left Eilís and Aoife, our new baby, at home. It was to be a foretaste of my life as a map-maker. The officers' mess in which I lived had its formal side. Dinners in full military dress uniform were a regular feature for all of us living there. It was the only time I wore my uniform.

I soon realised how formal these occasions could become. Eilís came from Ireland to attend the Christmas dinner. I was to pick her up at Heathrow airport early on the Friday of the event. But Heathrow was fogbound for most of the day and few flights were allowed to land. Eventually, Eilís's flight arrived, an hour and a half before the dinner was due to begin, and there was still a foggy M4 to Newbury to navigate in an ancient Austin A30 with very weak headlights. We were late. I expected that we might make it before the end of the main course. But no, everyone was seated in the dining room, probably had been for half an hour at least, waiting for us. We were the guests of honour and we did not know it – and the dinner could not start without us. I wanted to sink into the ground with embarrassment.

I noticed the differences in my new landscape as I practised field-surveying techniques in nearby farmland. Unlike Ireland, it was open and rolling. It was manicured, every inch used, with very little by way of scrubland. The hedgerows were maintained and evenly cut. The soil was lighter in colour, and the whitish chert stone that ran through it made it appear lighter still. But I could not sense warmth or depth in it as I did in the fields of Ireland. There, the fields and hedgerows preserved the traditions of generations. Here, it was as if the constant cultivating and manicuring had swept away all the stories and customs.

Nevertheless, small villages like Chieveley which were un-touched by developers' hands and where we went at weekends to eat, held the memories of the past. Red-brick houses, cottages with climbing roses and abundant flowers and pubs with low ceilings supported by old timber beams were from an era that had not caught up with time.

6

ALADDIN'S CAVE

I had spent a little time before I departed for England getting to know more about the modern mapping programmes to update towns and cities. My exploration of the new was short-lived for the past had quickly absorbed me.

I had been avoiding Captain Eoghan O'Regan, my commanding officer, an incessant talker always in search of a captive audience, when I strolled into an old outhouse workshop where a grey-haired man was repairing surveying instruments. I was curious about the instruments he was working on but more so about the piles of old copper plates stacked at one side of the workshop.

'They are the copper plates of the six-inch survey,' the man said casually. 'They're never used now.'

'Are they the original maps?' I enquired.

'They are. Take a look.'

In an instant I understood how the legacy of Thomas Colby had seduced the cartographers and surveyors who worked here. I too was seduced.

I had found the Aladdin's cave of maps – copper plates dulled from age and neglect with Ireland's history embedded in them. These were the original plates on which the first scientific and the most detailed survey ever undertaken of Ireland was engraved: seventeen hundred maps of all Ireland drawn in the twenty-five years before the famine of 1845. Here in this small, dusty, insignificant workshop I could travel through Ireland, mentally walk every road, visit every house, imagine the people and their customs, picture the seasons and the crops in the fields. Every iota of the landscape was meticulously cut into the copper: soft lines representing the Ireland of the past.

Placenames, townlands; bridges, streams, rivers; forests, bogs, fields – they were all part of the Ordnance Survey map, the record of nineteenth-century Ireland preserved right before my eyes. Ruined churches, graveyards, reiligíns; crannógs, stone circles, standing stones were all there. The surveyors and cartographers who made the maps had bequeathed a legacy far greater than a series of lines, symbols and names drawn on pieces of paper. They had woven a hidden code into every line they imprinted on the map, a code that could penetrate the psyche of readers and coax the spirits and presences that once inhabited the land back from their lifeless solitude.

It was strange to think that maps which were made for such functional and objectionable purposes – to impose a taxation system on the land of Ireland and fill the coffers of the treasury in England – could hold such magical power.

Not long after my discovery of these maps I travelled to County Cork with John, a surveyor, to become familiar with field-surveying methods. We called in at a farmhouse to deliver a map he had promised. It was one of those days when the sun shone from clear blue skies and the world was still except for the leaves of the trees gently swaying and whispering. Jamsie, the farmer, was at the door of the whitewashed cottage, clearly waiting for us.

'God bless ye, men. I thought ye weren't coming,' he said.

'Sorry, Jamsie, we were delayed on the road,' John replied. But don't worry. We have the map for you.'

The sun beamed through the open door and lit up the map as John spread it on the old wooden kitchen table. Jamsie couldn't wait to examine it. The map was a token of appreciation for him and his wife Agnes for allowing John to take measurements from the hill behind his house and especially for the feed of bacon and cabbage they had given him the previous week.

Agnes was puzzled. 'Where's our house? I can't make it out at all,' she said, as her eyes roamed over the fields and small boreens on the map.

'Look,' said John. 'Right here. See, yours is the long house and there opposite are the haybarns and the tractor shed. There's your boreen winding its way down to the main road.'

Jamsie took his bearings from the surveyor's finger and mentally entered the map. 'Agnes, there's the track leading to the well – and the well is even marked,' he said excitedly.

'Oh, look at all the twists and turns in that old track,' said Agnes.

'That must be Jack the boy Murphy's cottage,' Jamsie said, as his eyes moved along a narrow road in studied concentration. 'Christ! It's thirty-five years since he left for Boston. We never heard a word from him again.'

'Ah! Jack the boy,' sighed Agnes. 'I don't think he ever married. I wonder if he's alive or dead.'

Jamsie's eyes roamed across neighbouring houses as he recalled the living and the dead, the hard men of the drink, the women who were left on the shelf and the great firesides for stories and music. He spoke to no one in particular, more to himself.

'There's the townsland boundary running along by the stream in the lower field and turning down Conway's boreen. Look, even the areas of our fields are shown. There's the long field where the sycamore is; the map says there's ten acres and two roods in it.'

Agnes's eyes joined his as he walked along the roads, stopped at each house and lived his memory. She remembered the people too: her grandmother who lived in the next townland, her thatched house in ruins now. Nettles and brambles flourished where flowers had once bloomed in neat beds. And there was Mary Kelly's house next door.

'Poor old Mary,' said Agnes sorrowfully. 'She had the blackest head of hair you ever saw. God love her – she had a tough life minding that old father of hers when he took to the bed.'

Jamsie had known her too. 'He never gave her a moment's peace. Wouldn't let a man come near her. She never had a life of her own – poor oul crathur,' he said.

We listened in silence. Not wanting to interrupt the stream of memories flowing from the map, we crept over the worn limestone flagstones. We slipped through the open door into the bright

New Ross: extract from the Wexford six-inch Sheet 29, 1841.
(With permission Ordnance Survey Ireland)

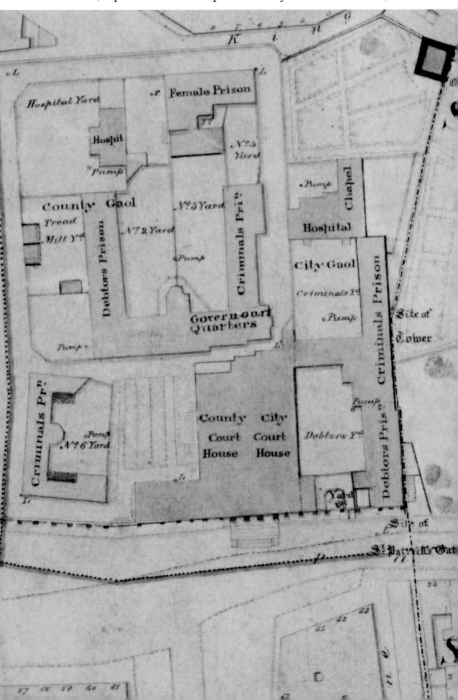

Poolbeg Lighthouse with the low-water mark,
engraving by Ordnance Survey circa 1837.
(With permission Ordnance Survey Ireland)

Benchmark at Cannon Street, Waterford.

Curraghmartin

Portnahully

School
94

R.C.Chapel Carrigeen

Lurranny
Smith

Corluddy
Castle

188 133

285
129

265
197
71

270 Smithy

76 Moonemn Lick**e**tstown

Wood
156

Moonveen

RIVER SUIR

GREAT SOUTHERN AND WESTERN (MALLOW AND WATERFORD BRANCH)

65

Gr

278

Corrisphierish

Carrigamore
Ho

Knock Ho.

Woodstown

Killoteran Ho.

Old Court
Church

Ballynane
Cemetery

Inn
Whelans Ho

Mount Congreve

227

Butlerstown
Hall

88
L.B.
Constab.y Ho
109

Evergreen
Cottage

oodvilla

amstown
76

Bawnfune

112

124
Whitfield
The Sweep

Butlerstown Castle

282
Ballycashin

Powersknock

Whitfield

298
R.C.Chapel

School

Lisnakill
Church

Kilronan

299

oughdineen
Castle

Smithy Shinganagh
Cross Roads

Pembrokestown

Sugar Loaf
Rock

Knockeen

282

Church

Cromlech

Pembrokestown Ho.

Extract from the Waterford one-inch map, Sheets 168 and 179, 1908.
(Reproduced with the permission of the Board of Trinity College Dublin)

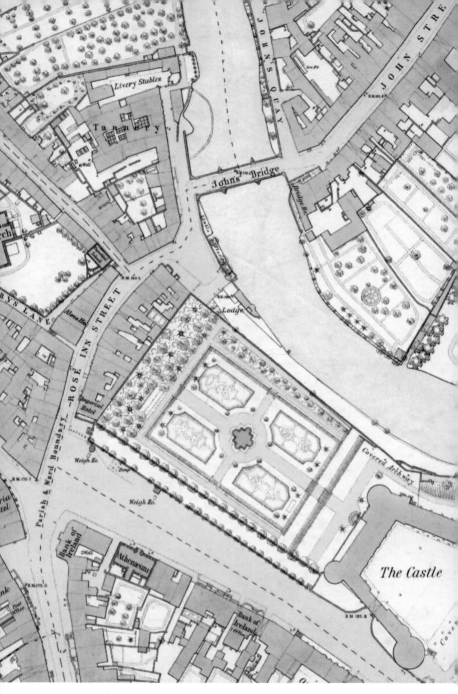

Extract from the five-foot Town Plan, Kilkenny, Sheet 19-47, 1871.
(Reproduced with the permission of the Board of Trinity College Dublin)

Part of the Discovery Series, Map Number 70, County Kerry.
(With permission Ordnance Survey Ireland)

Inishmurray, County Sligo.
(With permission Ordnance Survey Ireland)

sunlight and climbed the hill. Now we had another map to make, more angles and distances to measure.

After that, I took to visiting some of the men responsible for the various activities of the office. They elaborated on the stories Gerry Madden referred to on my first visit. Harry Crowe, a small, pleasant man eager to pass on his knowledge, was the superintendent in charge of stores. Late one afternoon he unlocked a heavy steel door and switched on the light of a bare bulb dangling from the ceiling.

'What is all that?' I asked, as I peered into the dimly lit room and saw nothing but old boxes, big, small, square, rectangular, on shelves and on the floor.

'That,' he said, pointing to a slender dust-entombed wooden box about ten feet long, 'is one of Colby's bars.'

'Colby's bars?' I repeated. 'What are they?'

'The Lough Foyle baseline was measured with them.'

'And what was that?' I asked impatiently.

'The very first line measured by Ordnance Survey in Ireland in the 1830s. And it was measured with these bars.'

Full of interest now, I drew nearer to inspect the box.

'Can you imagine,' Harry said, 'measuring nearly eight miles which included a river with those bars?'

'No,' I responded. 'And even if it could be measured, how accurate would it be?'

I was wrong. I found out that the bars were made from a combination of metals. Temperature variations had no effect on their lengths because of the way the metals were combined. Six bars measuring sixty feet in total were mounted horizontally in series on tripods and were gradually moved along the length of the line to be measured.

Harry told me that it took sixty men seventy days over a two-year period to complete the work. I was enthralled, imagining the work that had gone into measuring just eight miles. The sixty men worked summer and winter, dragged the bars across fields and bogs, meticulously set the bars up, waded waist-deep in the River Roe and at times waited impatiently for crops to be harvested. One officer proposed that Colby purchase the growing crops so that the men could proceed more rapidly with the measurement in fine weather. They took the measurements, then moved on and did it all over again and again. The bars had to be moved nearly seven hundred times as the men measured from start to finish.

'And you know,' Harry said, smiling, 'they got it right. When the line was re-measured electronically a few years ago there was only one inch in the difference. Doesn't that beat Banagher!'

He told me that Sir George Everest of Mount Everest fame was so impressed at the reliability of the bars that he used them in the Great Survey of India and that similar bars were used in the Cape of Good Hope.

Over the next few months, Harry and others filled me with other stories of the early years of Ordnance Survey in the nineteenth century. They relived every event as if they had been present: travelling the country, climbing mountains in wind and rain and sunshine, crossing rivers, hauling theodolites and tents up mountains. They might well have been part of it all.

Angles measured from the highest mountains in the dead of night; waiting patiently for a bright light to shine from another mountain, seventy miles away. Theodolites three feet in diameter; parts hauled up mountains in wooden boxes and assembled in howling gales, yet reading angles with the precision of microscopes.

I wanted to know more. I wanted to know the men who carried out this great work, whose pale busts lining the musty hallway to Gerry Madden's office showed nothing but their sadness. Colby, Drummond, Larcom – determined, passionate, creative men – did the impossible. They gathered two thousand one hundred men, mostly untrained, for this mapping adventure, for adventure it must have been, to climb every mountain, walk every road and field and capture it all on sheets of paper.

Thomas Colby, a man with only one hand, led the way. He climbed the mountains, the toughest of them, to train his men in measuring angles with theodolites: measurements that had to be perfect. He designed the great triangulation, the framework of triangles into which the country was divided in order to simplify the map-making. He pitched his tent and did mathematical computations on top of Slieve Snaght in the Inishowen Peninsula while waiting for the mist and rain to clear, risking his life. He was injured in a violent storm on that same mountain while a comrade was killed. It was not uncommon to wait a month or more for clouds to lift.

'Harry,' I said one day as I sat in his office thinking about the seventy-mile distances between the mountains. 'How did they manage to make their survey observations at night?'

He pointed to a strange outline sketch on a wall of the office, like something Michelangelo might have drawn.

'What's that?' I asked.

'That's the limelight, invented by Thomas Drummond.'

It was an astonishing invention – a tiny ball of lime lit by a mixture of oxygen and alcohol, giving out an intense light that could be seen up to seventy miles away. Nothing else could match it at the time. The limelight was designed principally to overcome the difficulties in observing at night the long distances between the Irish mountain-tops. It was also needed to take advantage of the tiny breaks that the misty and cloudy weather allowed on the mountains. But the weather had reigned supreme on these mountains since history began and it intended to show Drummond that he was at its mercy. On his first attempt to use the light, it forced him to sit out the entire month of November 1825 on a wet Slieve Snaght. Mist, that a light even of this intensity could not penetrate, shrouded the mountain-top. And in an act of supremacy, the elements sent a raging snowstorm that almost took Drummond, his helpers and the light to eternity. When the storm's rage was spent the light appeared out of the blackness of the night and was seen on Divis Mountain near Belfast, almost seventy miles away 'like a star of the first magnitude visible to the naked eye'. I wondered if Drummond and his men celebrated the event with a little of the alcohol from the light.

'You'd wonder how they communicated from one mountain to the other,' I said to Harry. 'How did one crew know when the other was finished observing the light?'

'I never thought about that,' replied Harry.

Just when I thought I had seen about everything I accidentally happened on maps that were etched on stone. I could not believe my eyes – tablets of white chalk-like stone with maps imprinted on them. There were about twenty abandoned in a corner of a store room. It would take at least two strong men to lift one.

'They're old printing stones. They were used as printing plates,' Harry Crowe said casually, as if everyone should know about them.

'When were they used?' I asked, as I examined the detail of the maps imprinted on them.

'I think they were used up to the 1950s,' replied Harry.

'How did they get the maps on to the stone?'

'You need to see the lads in Print,' he said. 'They'll know it all.'

I learned afterwards that these stones were used as printing plates for the one-inch series of maps. The stones could be sensitised with chemicals just like a piece of photographic paper and the image of the cartographer's drawing imprinted on them.

The 'one-inch map', as it was known, in which one inch on the paper represented one mile on the ground, was a more general map than the detailed six-inch series. A greater area of the landscape was shown on a smaller sheet of paper. Houses were dots; field patterns and townland boundaries were absent. The roads meandered from one crossroads to the next, one village to the next. Two hundred and five maps covered the country.

Each colour on the map was set out on a separate stone. The colours – brown-red roads, green forests, blue lakes and rivers – brought the map to life and helped to portray the three-dimensional nature of landscape. But most of all, the hachures or hill shading lifted the map from the page. The printers built the map, layer after layer and gave it an extraordinarily timeless appearance. Thomas Colby's successors had finished the first edition by 1858 and produced many variations of the map by 1913, including one showing the geology of Ireland.

There were only a few stones left. Some had being given as educational tools to colleges of art outside Ireland. By now the images of the maps had surely been scrubbed from them to make way for student exercises. The majority had found their way to army headquarters in Parkgate Street in Dublin to be trodden on. They were taken as slabs for a footpath. The stones that had contained the ancient history of the landscape disintegrated under the weight of trampling boots. The chalk and the images had dissolved in the rain and vanished into the earth. For the first time I felt pain at such desecration.

The 1906 coloured map I had had in my bedroom when I was a child must have been printed from one of these stones. I had not known its origin then. I knew nothing about printing stones or the Ordnance Survey but I had been fascinated by this map.

I set about searching the stones to see if that map of my childhood was there. And it was; Sheet Number 168, only the black outline of the roads and the physical features. The memories came flooding back. I had mentally roamed throughout this map in childhood, building up pictures of what the Waterford countryside might look like and what I might find there. These were the very same roads I had wandered with my father and grandfather, first on the bar of a bike and later on my own bicycle. I had superimposed my own mental images at locations on that map as we travelled. The seasons dictated many of those images. Spring at the Six Cross Roads when the blackthorn turned white with blooms and the birds were happy. Summer when I used to race along the cliffs of Dunmore East with my arms outstretched into the wind and listen to the echoing cries of the kittiwakes. Autumn when my hands were torn with thorns as I picked blackberries for my grandmother at Cleaboy. And winter was the time for coursing hares with greyhounds on the cold, wet fields of Gracedieu.

I had crammed other impressions into that map: wild flowers – primroses, bluebells, daisies: wild fragrances – rose, furze, wild woodbine: wild fruit – blackberries, crab-apples, damsons. The warm smell of summer growth in the dampness after a rain shower rested somewhere near Sugarloaf and the harsh smell of winter rain lashing bare hedgerows lodged on the Green Road near my home. And people were there too: the old men who walked the roads near the Holy Cross and raised their sticks in salute.

My father and grandfather sang out placenames on that map

as we travelled through the countryside with a respect beyond the ordinary. Towergare, Monamintra, Duagh, Fenor. It was as if they were paying homage to ancestors who once inhabited these townlands. The ancient name that adorned each townland set it apart from the rest of the landscape.

Each townland had a different significance for me. The townland of Knockeen possessed the magic of a druid's dolmen and beside it a graveyard with the names of those who inhabited the townlands in the eighteenth century inscribed on tombstones. Smoor was where my relatives farmed the land. Knockaderry had a lake which supplied drinking water to Waterford. It was also a place where my father often stood and leaned on the saddle of his bicycle endlessly chatting to a man in an old grey suit and cap about hurling, the weather and the crops and a place where I became impatient for him to move on.

I anchored snippets of Irish history and memories of my grandfather and father on the map in the place they were told to me even though they related to somewhere else. Every time I looked at Sporthouse on the map I remembered the tale my grandfather told me about the coffin on the ditch. He was an altar boy travelling with the parish priest to a house Mass in the early morning when they came upon two men carrying a coffin containing a body. They were resting and the coffin was perched on the ditch. Nobody else would bury the body because the family of the dead man had occupied land from which other tenants had been evicted. They had to bury him secretly at dawn. I never asked if the priest had said a prayer over the coffin. He told me it was not uncommon towards the end of the nineteenth century for people who occupied land in such circumstance to be boycotted. 'Your grandmother's people were evicted and nobody would buy an egg from the crowd that moved in,' he told me.

Riesk graveyard on the winding road from Dunhill to Waterford evoked another memory. We were passing late in the evening in the fading autumn light. My father related a story about his mother helping men on the run in 1921. She was a courier for the IRA and she had to bring money and supplies to the men. It was unsafe for men to undertake such operations as they were more likely to be searched than women. She was driven by pony and trap to the various hiding places, posing as a woman returning with goods from the market. She was never stopped or searched during her missions.

Town maps were not to be overshadowed by the six-inch and one-inch maps for they were masterpieces too. There were hand-coloured town plans of the 1840s which, apart from Dublin, were never published. I wondered how Ordnance Survey could have afforded such luxuries when so much other work was being undertaken. The buildings, gardens and streets were shown in the finest of detail. Maps of Maynooth and Monaghan and other towns even showed the windows and doors of public buildings. Every tree was accurately placed on the map.

Eleven maps covered my own city and I could walk through the streets of 1841 and let the maps tell me what life was like then. It was different; hundreds of water pumps, clay pipe and starch factories, salt works, bone yards, tanyards, malt stores, lime kilns and iron foundries. The city was alive with small industries.

There were streets and lanes that no longer exist, among them Garter Lane, Royal Oak Lane, Little Chapel Lane, Usher's Arch and Murphy's Lane off Patrick Street, the lane whose occupants in 1821 were, according to John O'Donovan, 'of the lowest class of tippers, whores and pick-pockets of the lowest most diabolical character'. They gave way to buildings and open spaces. And there were the streets that have long since changed their names; Queen Street and King Street became O'Connell Street, Little George's Street was now Gladstone Street, Nunnery Lane retained its religious connection but changed to Convent Hill and Fanning's Lane became Carrigeen Park. Formal gardens were plentiful as well as graveyards but neither modern maps nor the map in my childhood mind had any trace of them.

Harry and the other men brought the Ordnance Survey to life for me. I began to see the place and the buildings which had been so depressing to me in a new light. I filled the rooms with mental

images from the 1820s, images of the men who had sat on the
high stools and brought the island of Ireland into these rooms and
depicted its landscape in drawings. Each man had his particular role
in making the maps: engraving copper plates, calculating the areas
of every townland, determining the correct names for them, placing
symbols for trees and marshes and antiquities and finally printing
the finished maps.

Thomas Colby, Thomas Drummond and Thomas Larcom
became my heroes. The legacies of their achievements consumed
me. Their spirits were still in their maps and somehow they trans-
mitted to me their passion for the mapping of Ireland. I was now
ready to play my part in remapping the country.

I often wondered what Harry and the older men might have said
if I had unveiled the future of map-making to them. They would
scarcely have believed me if I told them that the air-survey machines
and their new drawing methods, introduced so recently, would be
obsolete in a few years. Harry might have had me committed to the
darkroom where Thomas Colby's bars were stored and thrown away
the keys if I had suggested that maps would be viewed on computer
screens and printed at the touch of a button. They would never in
their wildest fancies have believed that satellites and computers
would replace the traditions of one hundred and fifty years of map-
making before my career in Ordnance Survey came to an end.

7

Copper, Stone and Glass

I was curious about my one-inch map of County Waterford, Number 168, and wanted to find out how it changed from being an image engraved on a copper plate to an image on a stone, until it was finally portrayed on paper. I went to see 'the lads in Print', as Harry had suggested.

The reproduction department was a quaint, secretive place where a sort of brotherhood existed, preserving the past and sharing the secrets of the printing process among a chosen few. A code of honour existed through which the brotherhood protected its members. Anything that went wrong stayed within the limestone walls of the building. New recruits were only gradually initiated into the secrets, intricacies and rituals of the work. No one could come unannounced, not even the boss, as the print building had a commanding view over the other offices around the cobblestone square.

I climbed the steps of the fine limestone building which housed the reproduction department and entered a forgotten world from another century. The exterior gave no hint of the warren of old rooms, dark spaces, corridors, narrow stone stairs and vaulted basements within. It would not have surprised me to hear the outer door of the building slam shut and for some ghost-like figure to lead me through rooms where maps seemed to have been made for ever.

The heavy sounds of trundling printing presses and the smell of sweet, thick printing ink greeted me as I entered the printing room. The walls were a mosaic of faded green paint, daubs and splashes of printing ink and maps. There were two heavy black printing machines in the room and a smaller one in the corner.

Seán O'Carolan, the printing manager, began to explain, 'This is the flatbed printing machine. We print…

But I was lost, spirited back to the 1950s and my grandfather's printing works in Patrick Street in Waterford. I had escaped the hardships of school at lunchtime and raced the fifty metres down the hill to his workshop, which was called the Premier Press, a small inconspicuous two-storey building at the corner of the street.

My grandfather, with his collarless shirt, grey waistcoat, chain watch and inky hands, stood beside me in a white-washed printing room full of the sweet smell of printing ink. Dust, splashes of red and black ink and printed posters advertising dances long past adorned the walls. I helped him to operate the black flatbed printing press in the centre of the room, a trundling, great machine with huge ink rollers. We printed posters announcing dances with the Clipper Carlton and Mick Delahunty orchestras.

Another printing press stood in the corner of the room beside a dust-covered window. Here Bessie, my cousin, printed raffle tickets and memory cards. Bessie smiled whenever I came, greeted me with the same words every time – 'Well boy, anything strange?' – then carried on with her printing. The two machines mesmerised me as the rollers inked the printing plates and white paper turned to printed material.

Bessie and her printing press, which was like a huge sewing machine, intrigued my young mind most of all. I could never imagine one without the other. Up and down went Bessie's leg, working the foot pedal, all day every day, and a giant flywheel went clip-clop and kept the machine in motion. An ink roller click-clacked over and back and back and over, inking the typeface. Then a big paper holder dropped on to the type to accept its imprint. Up it came and Bessie's hands swapped white cards for printed ones. On and on went Bessie and the machine, working in complete unison. They were part of each other, neither ever making a mistake, always in synchronisation, Bessie like a gazelle moving ever gracefully in harmony with her machine.

Seán O'Carolan interrupted my reverie. '...This is the Furnival machine. It was used to print from stone. The one-inch map was printed here.' Suddenly I was back in the Ordnance Survey. This was the very machine my 1906 coloured map was printed on.

'Can it still print with stone?' I asked excitedly.

'Yes,' he replied in a matter-of-fact way and walked on towards the small machine in the corner. 'That's the original copperplate printing machine that was used to print the six-inch maps.'

What a treasure. This machine, about the size of Bessie's, had done something extraordinary: it had given the world impressions of the entire landscape of Ireland. The memories and stories embedded in the lines of the maps it printed still existed. The skills needed to print from it were still there, although now rarely used. I had to wait until 1985 to see an impression being printed from it: an engraving of Poolbeg Lighthouse, on the foundation stone of which the low-water reference mark for the heights of Ireland was carved.

I walked further into the building, into the womb of the reproduction department, where the cartographers' drawings began their journey of transformation into print. This was an ancient world of film processing and darkrooms lit by red safe lights. Mick Harford introduced me to something unique, a huge home-made cartographic camera to photograph or enlarge maps and drawings sometimes a metre in size. It was ten metres in length, and had been built around the turn of the twentieth century. The timber copy board that held the original drawing moved on iron rails so that the size of the enlargement could be adjusted. The interesting thing was that no room in the building was big enough to house the camera. The makers were resourceful; they placed the board which held the map to be photographed in a room which was at right angles to the board which held the film negative. The light that projected the map image on to the negative was bent through a ninety-degree prism.

'How do you light the map uniformly to ensure a good copy?' I asked Mick.

'No bother,' he replied. He took a four-inch nail from a shelf and tapped it into the centre of the map board. He lit four arc lights, two on each side of the board, moved them until there were four equal-sized shadows projecting from the nail, then smiled with satisfaction. 'See, it's easy.'

I met Mick again nearly twenty years after he retired. He spoke as if he was still there adjusting and photographing the maps.

'Do you remember the day you first came into the camera room?' he said with a glint in his eye. 'You were only a young fellow.'

'How could I ever forget such a place, with all its darkrooms and ancient equipment?'

'Well, I didn't tell you the half of it. Getting the correct-sized copies wasn't that easy. We had to do a bit of a fudge sometimes.'

He told me that the paper in older maps to be copied was often badly distorted, due to variations in its composition and temperature and humidity changes. No amount of adjustment of the camera would result in a correct size.

'We had a simple solution. We wet parts of the paper map and stretched it until the size was perfect. Needless to say, nobody ever admitted to such practices, but everyone knew.'

Nor did anyone admit to slicing slivers from the edges of distorted maps in order to butt-join them correctly.

Mick led me to rooms deeper in the building where doors had to be unlocked; the darkrooms where exposed film was developed – rooms that had never seen the light of day, with dim red safe lights, shadowy corners where ghosts of workers past might reside and oversee the work, plate-glass negatives, pegs dangling from wire lines, dripping film and baths full of chemicals. A man was washing film in a bath of chemicals, another cleaning a sheet of glass with methylated spirits. I was more than surprised when Mick told me that they used glass plates instead of film negatives. The days of plastic pre-sensitised film of this size had not yet arrived at Ordnance Survey.

'How do you make the glass light-sensitive?' I asked him.

'Look,' he said and took me into the next room, also shadowy and lit only by red safe lights.

Here, a man wearing a navy apron poured chemicals from open jugs onto a large rotating sheet of glass. I watched in awe as the liquid from each jug spread and settled evenly across the glass.

'It's all down to practice,' Mick said. 'You get a feel for it after a while. We do it with the printing plates as well.'

'What about the chemicals?' I asked.

'There's plenty of them,' Mick replied. 'Cyanide, silver nitrate, industrial alcohol, methylated spirits and a few more. No one was ever hurt.'

He told me they used the silver nitrate for sensitising the glass. A solution of cyanide potassium was used in the development process.

'You'd get splashes of silver nitrate on your hands and it would turn black when it was exposed to the light. The only thing to take it off was to wash your hands in the cyanide bath.'

On the way out I asked him about the great aroma of stew coming from the room beside the baths. He passed it off, saying it was Charlie having a cup of tea. I asked him again when we met twenty years after his retirement. He told me that there was a gas cooker in the room beside the baths of chemicals. Charlie the plate-maker was an ex-army cook and used to concoct stew there. He even added some of the industrial alcohol to it to give it a bite. It was hard to believe that no one was ever injured or poisoned.

'Ah, it was a great place to work in those days.' Mick told me. 'We all looked after one another. No one was ever caught slacking or doing nixers.'

There was quite a variety of items produced: election posters for Fine Gael, posters for St Patrick's Athletic soccer matches, music manuscripts and Christmas cards for boy scouts.

'We always knew when the bosses were coming and we made sure to be busy on map work. They never knew what was going on,' Mick told me.

I'd had my tour through the womb of the reproduction department and visited all the rooms through which my map of Waterford had passed. It was hard to believe that meticulous drawings and engravings such as my map and the six-inch copperplate maps could be dragged through such a cauldron of primitive chemical concoctions and emerge as perfectly printed maps. I had a new admiration for those cartographers, who for generations had mixed their magic potions and brought a paper map to life. Somewhere in this chemical process the memories of the land and its people were sucked from a great consciousness and indelibly imprinted into the lines of the map; memories like those Agnes and Jamsie had unravelled on their tour through the map of their townland.

More urgency crept into the print department at the beginning of the 1970s. The rate at which urban and rural maps were updated meant more printing. There was demand for more general maps of Ireland and local tourist maps as the tourist industry grew and the number of cars increased. Contract work, which had deadlines, was being undertaken; maps of Ireland had to be produced for oil companies and Bord Fáilte. Aerial photographs had to be processed for the air-survey machines and for customers.

In 1977, a dramatic change took place. Part of Gerry Madden's vision for a modernised Ordnance Survey came true and a new print building was complete. It was modern, built in brick and out of character with the architecture of the existing buildings: nothing more than a big warehouse.

The old methods and equipment were for the most part left behind. A new two-colour printing machine was installed. The copperplate printing machine and the stone-printing Furnival were relocated. The Furnival had been adapted to operate with modern printing plates. It was scrapped a few years later to make way for a further multicolour printing machine. It was too big to be accepted by any museum.

The new building was designed to be temperature- and humidity-controlled to ensure stability of printing paper and give more reliable printing results. State-of-the-art humidifiers were installed which were supposed to release a microscopic mist into the air and keep the humidity stable. But the microscopic mist turned out to be droplets of water which landed on the print paper. The humidifiers were eventually abandoned.

Locked doors were gone, replaced by light-proof revolving doors, and the darkrooms were now fully accessible to all. The stew room and the kettles were replaced by a bright canteen. Pre-sensitised plastic film and printing plates, already introduced on a partial basis in the old building, now completely replaced the old manually-made negatives and plates. Within another year or so, manual development of films was replaced by an automatic film-developing machine.

Cyanide was a thing of the past. Black silver nitrate spots on hands were no more. The old methods were gone; the new ones easier to learn. The intimate knowledge of traditions and skills perfected over a hundred and fifty years was lost in the space of a few years. Operations became functional rather than personal. Nobody lamented the passing of an old craft or wanted to retain relics from the past. The new would have its way.

The brotherhood was no longer in the ascendancy and bosses could more easily discover what was going on. But the code of honour – keep problems within the department – was retained. Twenty years after his retirement I asked Mick about this code of secrecy.

'I can tell you some things now that we're both retired,' he said with a smile. 'One night we were doing overtime to get a map of Killarney printed. There was a mad rush with it.'

The supervisor was supposed to be in attendance at all times but on this occasion he had gone off to something in town. One of the colour layers on a printing plate broke up during the print-run. It was not uncommon for such a thing to happen. The solution was simple: print the next colour and get a new plate made the next morning.

'First thing next morning the supervisor got Charlie the plate-maker to make a new one. It was back on the printing machine before ten,' Mick continued. 'But the bosses found out and there was war.'

'What was the big deal?' I asked. 'There was no loss of time, was there?'

'No, but some spy told them that the supervisor was missing and they made a big thing out of it, although they could not accuse him of being absent because everyone else would swear he was there.'

He told me that they came storming into the print room demanding to see what went wrong. They could find nothing because the plate was back on the printing machine and that particular colour was being printed.

'Why were they so excited?' I asked.

'Ah, you know in those times everything was a big deal. They really wanted an excuse to catch the supervisor.'

There was an internal enquiry to determine who the tell-tale was. The suspect was called to the supervisor's office. The door was firmly closed behind him.

Mick concluded, 'Everyone knew what happened. Every sound could be heard through those walls: raised voices, furniture shifting, a thud and someone falling against a wall. But it was never discussed.'

By the late 1970s the changes in Ordnance Survey were perceptible. The new print building was symptomatic. There was an urgent need to produce more maps and bring the old ones up to date. Money had become important and a greater percentage of costs had to be recovered. The print department had for many years engaged in outside contracting: they had had little competition but now the outside companies wanted better value and were inviting competitive tenders. It was a struggle for Ordnance Survey to compete.

8

INTO THE FIELD

The buildings had not changed when I returned from England in 1973. More intelligent spiders had sought new, inaccessible places to make their webs. Thomas Colby's bust in the hallway continued to ponder the fate of his records in the musty basement. Gerry Madden was no longer there. He had died suddenly the previous year.

However, his legacy remained and his plans for the revival of the Ordnance Survey and the renewal of the maps were taking shape. The long tradition of an all-male map-making empire had been swept away. Civilian female employees had passed through the heavy green gates and taken their places in the drawing office. The monastic ambience of that office would never be the same again. There were five females at first, all holding the grade of mapping draftsman. They were not treated the same as the men. Their duties were confined to drawing and their salaries corresponded to their single-duty status. But it was a beginning.

Soon the floodgates opened and, throughout the 1970s, further civilian cartographers were recruited. The first were employed in 1973: boys and girls fresh from school. Long-haired men and mini-skirted women came to mingle with short-haired soldiers: new, bright colours infused into an older world. It seemed an odd mix at first, a culture where discipline was paramount mingling with another born in the freedom of the 1960s. Over two hundred school leavers joined the Ordnance Survey, more than doubling the numbers employed. They brought new vibrancy and an expectation that there would be a revolution in map-making in Ireland. At last,

Ordnance Survey had the wherewithal to bring the urban and rural maps back to a standard of which Thomas Colby would have been proud.

The training school was expanding in 1973 to cater for the intake of new recruits: two full classes a year instead of the usual handful of soldiers. I was sure my first job would be as an instructor there but it was not what I wanted. I was not attracted to the prospect of being trapped in the old wartime felt-roofed wooden huts which housed the school. It was not my idea of a map-maker's job and I wanted to be in the open countryside. I'd had the taste of mapping the countryside in England and I wanted more. Happily for me, Eoghan O'Regan, my commanding officer, was content to instruct in the school and did little by way of other mapping work.

In any event, Muiris Walsh had other plans for me. Muiris, who had been Gerry's deputy, was now in charge. He too had been a commandant in the Corps of Engineers. We didn't know each other very well and hadn't met that often before I went to England, as he spent a considerable amount of time in the field. We officially met in his sun-filled wood-panelled office in Mountjoy House, which overlooked the lawn. Shelves, laden with books of records, hid behind the panels. The office had been the library of the old house and was now the finest office in the building. It retained the ambience, the warmth and opulence of an old library stocked with the learning of centuries. There was a strong aroma of stewed coffee coming from a bubbling and gurgling silver percolator.

Muiris was a friendly man who liked the informal approach. He conducted his meetings in low armchairs around a glass-topped coffee table, offering thick black coffee in white bone-china cups to visitors. I soon discovered that he did not just work there. He *was* the Ordnance Survey. It lived in his psyche.

'You've been to the Survey School in Hermitage and you know the theory but there's more to mapping than that,' he said.

I looked at him quizzically, not knowing what to expect, and replied, 'I hope you're going to send me to the field. I'd like to get stuck into the work there and get some practical experience. I'd like to see how the field sections operate.'

He looked me straight in the eye. 'The field surveyors have real experience. Learn from them and you won't go wrong. But you must earn their respect by showing that you can manage the work and get things done.'

That was the first and last lecture I ever got from him. We sat for a while drinking black coffee while he told me about his yacht in Howth Harbour. A crane would lift it from the water early on Saturday before the winter storms came. After that he would play golf at his Royal Dublin Club. Very interesting but I wanted to get back to the business of Ordnance Survey.

'Why did you need to do a complete retriangulation of the country?' I asked. 'Was Colby's original triangulation not good enough?'

'We knew that there were inconsistencies in Colby's original triangulation and it would have been unwise to use it as the basis for modern mapping. The coordinates of some triangulation stations like Hungry Hill in Kerry were not correct.'

'It must have been tough, especially with all those pillars you built on top of mountains.'

'Tough but fun,' he responded.

The job had been slightly easier than in Colby's time. More compact theodolites, better observation lights, the advent of electronic distance-measuring equipment and motor transport made things more manageable. There were still huge difficulties. Concrete pillars about three feet high, known as trig pillars, marked the apexes of the triangles. The new pillars were built to ensure that observation instruments could be mounted on a stable base and not be blown about in storms.

Originally, small triangles carved on rock or buried red tiles marked the apexes. The pillars, like the original survey marks, were built principally on inaccessible mountain-tops. Observations were still carried out mostly at night. Surveyors still had to climb mountains and haul a ton and a half of gravel, cement, water and heavy wooden casings up to the summits by hand. They still had to endure mists, storms and long waits in the cold. Wet tents continued to be part of the job. During storms, John Danaher, the principal observer on the scheme, tied himself to the trig pillar on Mount Errigal in Donegal and other pillars on the unforgiving mountains of the west of Ireland.

'It took three years from 1962 to complete the job,' Muiris said. 'I travelled over a hundred thousand miles a year and changed my Volkswagen every January. I was hardly ever at home.'

He told me that he often climbed two mountains at night, especially if there were difficulties with the observations. Sometimes he had no recollections of the roads he travelled. He fell asleep at the wheel one night, overshot a bend in the road and found himself parked in a field when he woke at dawn the following the morning. Luckily, the gate had been open. That day he had travelled over two hundred miles in the west and climbed two difficult mountains.

'It was an all-Ireland scheme. Did you have any problems with that?' I asked.

'We have a great relationship with the Ordnance Survey of Northern Ireland. But the B-specials, they were something else.'

Despite the political division of the country, the triangulation was carried out jointly by the two Ordnance Surveys in the border areas. It was dangerous, especially at night, to travel near the border along meandering roads that didn't care which side of the border they were on. The IRA was active at that time. Members of the Ulster Special Constabulary, B-specials as they were called, set up checkpoints, often near bends in roads where they would not easily be seen.

Muiris said, 'It was scary driving along narrow roads in the black of night with southern-registered cars. The potential consequence of driving through one of those checkpoints was unthinkable. We were regularly interrogated and searched.'

Muiris asked me to manage a triangulation project in Munster and south Leinster. He said he was a bit worried about the speed of the work but did not elaborate any further. He could not spend as much time in the field since he had taken over from Gerry Madden.

No one briefed me on how the work in Munster was going. John O'Hagan, the man directly responsible for field surveys, just gave me a map of the working area and the names of the surveyors. It was a little like learning about the history of the Ordnance Survey before I went to England. I picked up snippets as I went from office to office but no one ever explained the overall picture. There appeared to be an assumption that everybody automatically understood what was going on.

Needless to say I was delighted at the idea of working in the countryside, particularly as part of the work was in my own county. The countryside was where I always felt contented and free and it was a place of happy memories. In the countryside I had cycled and listened to stories told by my father and grandfather. I had already decided that office life was not for me. That eight-week spell working as a design engineer in the ESB headquarters in Fitzwilliam Square made my mind up about that. I had vowed with every steel beam I selected for Tarbert Power Station from a boring book of tables that I would not live my life this way. This was where I was meant to be.

I voyaged into the unknown in this, my first field job. I was an army officer and did not have direct responsibility for the people I was to manage. Yet I was responsible for getting the job done. There was a peculiar organisational set-up in Ordnance Survey. Soldiers and civil servants worked side by side. The soldiers, once trained, were assigned to civilian bosses and were responsible to them for their daily work. Army officers, like my commanding officer and I, were employed as professional advisers and had no direct responsibility in the management of the organisation. Yet we were responsible for organising and successfully completing major projects like the triangulation of Munster.

I met Jim, the supervisor of the field section, outside the Bridge Hotel in Waterford on a grey, wet November morning. It was our first meeting and neither of us knew what to expect of the other. There was a reconnaissance to be carried out to find suitable hills for pillar-building and observations. We needed to get permission from farmers to use the land around Mullinahone and Slievenamon in south Tipperary. The field crews were building pillars elsewhere.

I was surprised to find Jim immaculately dressed in starched shirt, bright yellow tie, sports jacket, pressed trousers and shining brown shoes. These were not clothes for the hills. His use of Imperial Leather aftershave was liberal. I assumed that the intention was to let me know who really was in charge. He was a smooth, well-spoken man with a sincere, convincing voice. But he became a little flustered when I enquired about the technicalities of the work.

He wanted to concentrate on the lower hills first. It suited me as the weather was bad and I felt we could at least achieve something during dry periods. We climbed through wet fields and furze, mud and rock to the hilltops. Jim added a pair of green wellingtons to his immaculate apparel and these made him look like a squire and me like his helper. We sat in the car when it was raining and watched the grey, full clouds dump their load while lonely crows flew for shelter.

Each day Jim arrived at our meeting place impeccably dressed. He did not eat much during the day. His lunch was just two hard-boiled eggs wrapped in tinfoil. 'I always carry two hard-boilers in case I get caught on the mountains,' he told me.

We stayed for the first few nights in well-worn local hotels. Jim ate little and when I offered him a drink protested that he rarely touched a drop. When he got to know me a little better he asked if he could return to his home sixty miles away each night as, 'The wife is not too well and is on her own.' I had no real objection as long as we met early each morning and got to work.

The arrangement worked for a day or two but after that his punctuality deteriorated and with each day the smell of aftershave got stronger. He was convincing in his reasons for the late arrivals. 'The wife was really bad last night; I got a bloody puncture and had to change the tyre; the engine broke down; the car wouldn't start; the fan belt was slipping.' There was no shortage of excuses; no bounds to his apparent embarrassment and remorse at being late. Yet there was never a dirty hand or a smudged shirt to indicate that there had been a problem with the car. I insisted on getting the work done regardless, climbing hills late into the evenings until it was dark, then planning for the next day before he left for home. He realised I was becoming impatient but his pattern of late arrivals continued.

It is strange how the intolerant expect forbearance. Jim's field teams worked hard and were experienced surveyors. They knew the countryside; they were part of it. Time was rarely a factor when a job had to be done on the hills. Fair weather was used to the full. The quid pro quo was that the teams could take long weekends or time off when a job finished.

Jim and I finished the reconnaissance and joined the field teams for the observations on Slievenamon and the Comeragh mountains. Monday was normally a late start and Jim, who had been punctual on this occasion, paced the pub car park where we were to meet, impatient for the last of the surveyors to turn up. He was indignant by the time the man arrived. It was time for him to test my authority. He rushed in without asking the man's reason for being late. As it transpired, he had a genuine family problem.

'Captain Kirwan,' he said, (I had recently been promoted), 'this is disgraceful and unacceptable. I recommend that you return this man to the office.'

I said nothing.

We negotiated the long incline to Seefin in the Comeraghs and measured the distances to Slievenamon, Galteemore Mountain and some lower hills, then moved on to the next observation station. We worked late into the sunset to complete the observations.

I did not send the man to the office. Jim's punctuality deteriorated further and at times he did not turn up. My patience gone, I returned *him* to the office. John O'Hagan was surprised and taken aback by my actions. After all Jim was a senior and experienced man on the field. I had been naïve. Everyone except me knew about Jim's drinking habits. I was the first to address them, even if it took time for them to register with me. I learned then what Muiris Walsh had left unsaid when he asked me to undertake the job. Perhaps he was testing me.

I loved working around the country. I was always happy to leave Dublin early on Monday morning and head for another Ireland. There, life was lived at an easier pace and everything was in tune with nature's time: a chat with a farmer about the weather and the crops or Sunday's match; a cup of tea from the woman of the house or an explanation of what we surveyors were doing.

There was predictability even about traffic on the roads,

something that on occasion nearly caused me grief. Locals travelled at the same times each day to the creamery, the post office and the schools and knew what other traffic to expect. They were wont to take the full width of the road and an unexpected car or Ordnance Survey van weaving around narrow corners often caused panic. 'Nobody ever comes along this road at this time,' was inevitably the first phrase from a local's mouth. But time could then stand still for a leisurely conversation about maps and the post office or creamery would be forgotten.

There was always something new to learn, even the obvious that I had never noticed before. One day I stopped at a cottage where an old man smoking his pipe was leaning on the gate.

'Good morning,' I said. 'Could you tell me the best way to get to the top of Tory Hill?'

'God bless you,' he replied. 'It's that time of year again.' He lifted his arm slowly and pointed to the two rows of swallows on the nearby electricity lines. 'They're getting ready to go.'

He continued to lean on the gate watching the swallows' every move and through his silence invited me to join him. There were hundreds of swallows, tightly packed, side by side on the wires, chattering, shuffling, flapping wings, exchanging glances. Perhaps there were words of comfort to a loved one or advice to a youngster about to take a first flight south. Or perhaps it was excited chatter about what lay ahead at their destination.

Every now and then a swallow fluttered from its perch and settled among others in the line. There had to be a story or a piece of gossip that needed telling. More flutters and groups of five or six left the line and swooped into hawthorn bushes. Was it for private conferences, secret words of advice, or to pass on some ancient memories within a family? Their business done, they returned to the line.

The aloof, the lonely, maybe the ostracised, perched singly away from the rest. Was it quiet contemplation before the flight, worries that an offspring would not survive the long journey, or was it the knowledge that perhaps this would be their own last flight?

Suddenly, their business done, the swallows flew from the wires in great circles over the fields, darkening with their shadows the place where we stood. They returned to the wires: more chatter, flapping of wings, fluttering, more questions, advice, reassurance. The aloof returned to their lonely perches. Five or six times, the swallows made this flight in ever greater circles. Finally they were gone.

We stood for a few more minutes in silence. Had an hour passed?

The man took me by the arm and said, 'Is it Tory Hill you're looking for?'

It was easy to work on the lower hills, where a van or a tractor could haul building materials and equipment. The mountains were less accessible and rarely welcoming, setting obstacles in our paths to the summits; long waterlogged approaches from the roads, bog holes, heather and thick energy-sapping grass, rocks, sheer faces and running streams. But the rewards on these summits made the climb worthwhile.

I found vistas new to me and often unexpected. I could tune into the different moods of the surrounding landscape: the light and freshness of lush, open countryside with crops and beefy cattle; the cold and damp of wet lands where grass grew reluctantly and hungry cattle stared in misery; the thorny briars and furze of rocky lands; the craggy coastlines and the islands like pieces of rock tossed randomly from the mainland. The mountains revealed their ever-changing moods: happy in the sunlight, mischievous in the rain and fog, haunting when cloud shadows raced over them and flocks of sheep ran before a farmer's dog.

But nature showed me early on that it was master of all I surveyed. I stood in the sunlight one morning on Slievenamon, a mountain I had climbed many times, measuring the distances to the surrounding trig pillars. By early afternoon a mist had descended and rendered my sense of direction useless. I was imprisoned within the untouchable boundaries of an endless mist and could do nothing but wait for nature to release me. It did not help me to curse my bad luck that I had not left earlier or fret that I might not complete the week's work on time. I might have learned that day that plans cannot be forced but this lesson was a long way off. Here the stones and rocks, the heather and grass rested in the knowledge that sunlight would return in its own good time. They had time to wait, time to accept the mist, time to be. I did not.

Winter could be particularly hard on fieldwork. None of us minded too much; we were young. Rain and wind, mist and penetrating dampness were constant companions, often delaying the work for days. Cold nights in old hotels, wind rattling through windows, bed linen that might not have been changed, rust-stained and mouldy bathrooms; musty dining rooms with faded embossed wallpaper, leathery steak with greasy onions, infernal mixed grills steeped in lard, tea with stale bread and butter. A soak for an hour in a hot bath was the only consolation.

Winter also brought long boring days sitting in Land Rovers waiting for a break in the clouds. I arrived in Belmullet, County Mayo, late on a Thursday afternoon in November to visit Harry McDermott and his field section. Harry, a patient man, had sat for the week watching the storms blow in from the Atlantic. He had not measured one line during that time. Next morning at breakfast, I could see that he was edgy, watching the rain lash against the window.

'I'm afraid there'll be no work today either,' he said with a resigned sigh. He turned to Malachy McVeigh, a member of the field section. 'Mr McVeigh, how long more is this rain going to last in this God-forsaken place?'

'It has to break soon or we'll all go mad,' Malachy replied. 'I've never seen it so bad.'

Harry turned to me and in a slow deliberate voice said, 'You know, Captain Kirwan, the British didn't beat the French here in 1798. They were all drowned in the bogs.'

It could be like that in the depths of winter. But Harry and Malachy soon settled down to another pot of tea, a slight nod and a hand on the teapot drawing the landlady's attention. They had learned patience. I might have taken a leaf from their book.

'Mr McVeigh, it's a different Ireland here,' said Harry. 'What did you think of that little lad doing his homework yesterday?'

'I don't know how he does it. He'll be blind before he's twenty with that light.'

'One bare bulb for that big kitchen. It couldn't be any more than twenty watts and the mother trying to darn socks with that light as well,' replied Harry.

'I suppose they're lucky to have electricity at all in an isolated place like this. It's probably not too long ago that they had only candles,' said Malachy.

'Well they certainly don't expect as much as us living in Dublin. My house is lit up like the *Innisfallen* sailing down the Lee at night.'

Harry told me that they had sat in a Land Rover at the foot of a mountain the previous day hoping for a break in the skies. The weather forecast was wrong; the clearance had not come. A woman living at the foot of a mountain had taken pity on them and invited them in for a cup of tea.

'There are some very generous people around, Captain Kirwan, especially those with little,' he said.

I saw it elsewhere on my travels in the 1970s, an aspect of an Ireland content with what it had. A whitewashed house near Malin Head had tiny windows, an open door its only source of light. Electricity had not penetrated this house, nor had running water or an inside toilet. An old man sitting in a smoky room lit by a turf fire, smoke ingrained into the wrinkles of his hands and face, was happy to greet me. And in Wexford, too, a farmer with a fertile holding living alone was happy to welcome me by candlelight. 'I'm happy with what I have,' he said. The black kettle was boiling on the open fire, a blackened teapot and two cups on the table nearby.

Fieldwork was not all about climbing mountains. Surveyors had to establish the height above mean sea level of points along the road network of the country: this work was called levelling. It was a repeat of what was done in the 1830s except that then the surveyors were measuring the height above low-water mark at Poolbeg lighthouse. Now the surveyors were measuring the height above mean sea level at Malin Head. Mean sea level had become an international standard for measuring height. There was a difference of more than two-and-a-half metres between the two, something which caused endless confusion for engineers. This was especially the case if they were using a mixture of old and new maps. One might contain mean sea-level heights, the other low-water mark heights. There was a further complication for engineers working along the border. The Northern Ireland height system was based on mean sea level measured at Belfast Lough. There was a difference of one tenth of a metre between it and Malin Head.

It was a slow job. The surveyor walked the roads to rediscover the original benchmarks for which precise heights were known. A benchmark, often called a crow's foot or sapper's mark, is a three-legged mark topped by a horizontal bar and carved on rocks, gate pillars or the cornerstones of houses. Some were easy to find, especially the ones on gate pillars and houses. More often they were hidden under vegetation gone wild. Many had disappeared, having been removed during road-widening works.

As I walked these roads I could not fail to remember the Sunday mornings of my childhood, when I cycled the narrow roads of Toureen and Monamintra with my father and grandfather. The road boundaries were probably just lines of loosely laid stones when the sappers passed by. But as we passed, the stones were barely visible. Nature had endowed them with coverings of soil and rough grass, brambles, dandelions, buttercups, daisies and little blue cornflowers, colourful and happy in summer, wet and miserable in winter. In other places briars, hawthorn, blackthorn and furze hid the stones from view. Some stones, hidden entirely from sunlight, which had managed to escape and retain their independence for a time, were smothered by blankets of soft wet moss. Others that regularly took the light and heat of the sun were spotted with map-like patches of thin lichens. I wondered now whether any of these stones bore the imprint of a sapper's chisel and whether any of the men who sat outside the cottages on those roads had known that the sappers had worked along those roads.

Perhaps one of the men had salvaged a stone with a crow's foot and stored it in his barn to wait for the return of a surveyor. This was not uncommon. In my life as a map-maker I found that rural people often had an inordinate respect for the crow's foot. Frequently, when roads were widened or sharp bends removed, a local man would salvage the mark and keep it safely in a barn. I was sometimes taken to the barn to see it. Or maybe one of those men sitting outside his house had repositioned a mark when he himself had moved a gate pillar, like the old man I met when I was working with the levellers. He told me that he had taken care of the crow's foot. The sappers had asked his grandfather to keep it in good order. Each generation took good care of it. But a few years previously this man needed to widen his laneway. He had rebuilt the pillar in a new position and carefully replaced the stone bearing the mark. He did not realise that this was worse than removing it completely – that it was the worst possible thing that could happen, especially if nobody admitted to doing it. It was the original position that was important and the benchmark could never be restored to its exact height above sea level.

The new marks placed by Ordnance Survey in the 1970s were very different from the finely-carved crow's foot of the sappers. They were just unattractive and insignificant steel bolts driven into a wall of a house or a footpath. They were meant to be stable, a constant reference to the height of that place above mean sea level. I remember being horrified when I first discovered that these marks were often placed in footpaths surrounding new houses. This was done for convenience when a house owner didn't want a steel bolt driven into the gable wall. My experience as an engineer was that footpaths were notorious for subsiding, as many of them had poor foundations.

The new marks were purely functional. There was little hope that they could ever generate the sort of folklore associated with the original marks. Ireland was on the move: there was no longer time to savour the aesthetics of a crow's foot, no longer time to take time.

9

THE BOUNDARIES OF MAN

It did not take long before the map and my surroundings were one. I could not wander or travel in the countryside and not mentally map it; I could not walk the streets of a town without drawing its shape in my mind. I became as perceptive of my surroundings as I had been in my childhood.

I wondered why fields were shaped in the way they were. They could be small, big, long, regular and irregular, their boundaries often following the courses of streams. Who built all the boundaries?

I wondered how or why small roads meandered the way they did, twisting aimlessly through the countryside, never taking the shortest way between villages or towns. But then, it would have been boring if you were sure where each one led. There was mystery and uncertainty around every twist and turn. There was a choice to be made at crossroads without signposts, frustrating when time was against you but a reason to explore where each one led when you were not in a hurry. Why did the towns on the borders of Kildare and Meath assume their unique shapes and characters; Kilcock, Enfield, Johnstown Bridge, Carbury? They were all different, recognisably individual on the map, set apart by the distinctive lay-out of their streets.

And the townland boundaries? For the most part there seemed neither rhyme nor reason to their locations. Only Richard Griffith and the men who marked out the ancient divisions in the 1820s knew why a townland boundary separated one field from the next, or made one person different from another because they were born on the opposite sides of a boundary.

Eilís and I and our two daughters settled in Nicholastown on the borders of Kildare and Meath, twenty miles from the Ordnance Survey office. Part of the townland poked unnaturally into the adjoining one, Killickaweeny. It was as if Griffith or his men had acted out of perversity or had settled some ancient feud. Nicholastown itself was a nondescript townland, its only boasts having the house of the landlord's herd or cattle drover on it and being mostly fertile land.

That was until the 1940s, when the Land Commission acquired and subdivided the Colgan estate and introduced farmers from the west of Ireland. It was a far cry from when the Empress of Austria hunted here in 1879 with the landed gentry. Now a new culture and lifestyle entered the old townland. Mrs Daly, her husband, 'the Boss', a gentleman farmer and part-time auctioneer, and her eight children had given up their farm near Belcarra in County Mayo for better land here. Séamus, a younger son, ran the farm.

We regarded Mrs Daly as the dowager of the community, a font of wisdom and knowledge, well read and versed in current affairs, history, folklore, the church and herbal medicines. She was a practitioner in herbal remedies long before they became popular, to the extent of growing her own wheatgrass. She visited us during the bright evenings and spent hours sitting in our kitchen in a bentwood armchair with both hands solidly placed on her walking stick. She was like a grandmother to us. The new world of the 1970s alarmed her. After we had considered the affairs of the world together, she always ended the conversation with, 'We're living in quare times. I don't know where it will all end.'

Dónal Fenton, another neighbour, came from County Kerry. A confirmed bachelor, he minded his demanding widowed mother, a retired schoolteacher. The Land Commission had also resettled the herd from the original estate and his son, Paddy Bowe. They now lived in a Land Commission house on thirty acres. Paddy's wife came from Kerry, having worked in the munitions factories in England during the Second World War and as a servant in big houses in Dublin. She happened on an advertisement which Paddy had placed in the evening papers and met him in Barry's Hotel in Dublin. They married soon afterwards. She waited hand and foot on him and prayed that he would be in good spirits when he returned from jobbing on cattle lorries full of stout. Paddy spent much of his time grazing cattle on the long acre or chasing the ones that escaped from his fields. He always saw the bad side of the weather. It was always a 'miserable day' in the winter and 'Too hot. 'Twould kill you,' in the summer.

We were the adopted family of that migrant community who were all part of Nicholastown now but whose customs and traditions and souls remained in their native counties, Mayo and Kerry. Among themselves they were still living on the western seaboard. Theirs was a Gaelic world filled with the lore and mythology of a bygone era. They returned to the west at every opportunity, in memory at least, to relive the lives of people who had long since moved on. Mrs Bowe and Mrs Fenton and other migrant farmers' wives in neighbouring townlands relived the old days in their native Irish tongue. It was easier for them.

Like the townlands left behind in Kerry and Mayo, the townlands in Kildare and Meath took on a mystical significance to those who left them. The old bog road in Killbrook, where hot springs flowed from the earth, was immortalised in song by one of its emigrant natives, Teresa Brayton, as was Ferns, in whose church her mother's funeral Mass was read, although the old bog road was probably a miserable place to live in during the mid-nineteenth century.

Three townland boundaries crossed the narrow roads where we lived in Nicholastown. We could not leave the house without entering another townland. In Killeighter, an old graveyard stood aloof from the nearby road which led to Kinnegad and Killucan, where we bought second-hand furniture from antique dealers at weekends. It was inexpensive and we could not afford good new furniture. Boycetown came between us and the main Dublin-Galway road, the road to the Ordnance Survey. County Meath was less than three hundred metres from our house, yet we had to cross the townland boundary of Farrens, or Ferns as it was locally known, to get there. I crossed these invisible divisions on the roadway each day.

Few passers-by would be aware of the old boundaries I crossed on summer evenings when I strolled to the Rye river with Eilís, Aoife and Mary, our toddler, and sometimes with Mrs Daly as well, past the wooden stand where the milk churns stood, full in the mornings and empty in the evenings, and on which the children sometimes danced a jig or reel or recited a poem. The Rye here was no more than a stream flowing harmlessly over stones and pebbles towards Maynooth a few miles away, where its waters would be polluted – a sign of the modern Ireland on the move.

This river was the boundary that Griffith deemed should divide Meath from Kildare and the townland of Nicholastown from Farrens. The boundary was to be the centre of this stream. But this harmless stream had other ideas; over the century-and-a-half since Griffith defined it as a boundary it had changed its course and meandered its way through Meath and Kildare, leaving the boundary behind. This was the fate of many townland boundaries, clearly defined as the centre of a stream or river on the original maps, now an invisible line running through a field because a waterway had gone its own course.

We might have been happy living in Nicholastown for the rest of our lives. But just as we fully settled and became part of the greater family of the migrant people from Kerry and Mayo we suddenly decided to leave the area. We had been unable to get our eldest child into the local school in Kilcock because the admission age was raised from four to five. We were enthusiastic parents and believed that our daughter was more than fit for school. The parish priest was more interested in admiring the blackberries on the bushes than keeping us in the parish. 'Aren't the blackberries beautiful?' he said, when we met him walking on our narrow road in Nicholastown and broached the subject with him. In any event I was finding the twenty-mile journey to work through Kilcock, Maynooth and Leixlip too time-consuming: forty-five minutes each way in 1977.

We left the Rye to continue its meanderings and steal into Meath and back into Kildare. We left the old bog road and our migrant

family and headed to a townland called Commons on the borders
of Dublin and Kildare. People called the area Hazelhatch, the name
of the neighbouring townland and of a disused railway station
nearby. It sounded better. Commons was a townland that had
no history, or if it had, nobody knew it or was interested in it. We
bought an abandoned, partly-built house and I set about completing
it myself in the summer months. I moved into quarters in the
Ordnance Survey and Eilís and the children went to her parents in
Kilkenny. Working by day in the Ordnance Survey, by night on the
construction of the house and making a few visits to the field did not
leave much time for anything else.

Living here was not the same as Nicholastown. We were
newcomers in an established community and the closeness of
our wider family in Nicholastown did not exist. The names of
the townlands – Commons, Hazelhatch, Colganstown, Balscott,
Commons Little – did not have the warmth and lyrical tones of the
anglicised Irish placenames in Kildare. Strangely we were at another
boundary. Hazelhatch met Balscott, a tiny townland of nine acres,
on the border between Dublin and Kildare. Our house was where
Commons met Colganstown. The boundary between the two
townlands floated somewhere in the stream behind our house.
The twenty-five-inch map said it was six feet from the face of the
fence, but where was the face of this fence in the late 1800s when the
map was made? Whenever I worked in the garden near the stream
I pondered where that boundary might be. I liked to think about
that: it always held some sort of mystery for me. It had been the
boundary between the landed gentry of Colganstown and the poor
of Ireland. Why then was it not clearly defined? Was our townland
a commonage because the land was so bad that the landowners of
Colganstown did not want it and gave it to the poor? The earliest six-
inch map showed two houses in this townland but they were long
gone. The purple-yellow soil could not have sustained many crops:
it was hard and caked dry in the summer and muddy marl in the
winter. My attempts to grow vegetables failed miserably.

The surrounding countryside was flat and uninteresting. The
Grand Canal flowed through the townlands, its monotony broken

only by the lock gates near Lyons Estate and the narrow humped-
back bridges spanning it. The bridges were built by Lord Cloncurry,
the landlord of the townland of Lyons. They were extraordinarily
narrow, barely wide enough to accommodate his carriage. Local
legend has it that the narrowness was a consequence of the
landlord's meanness.

The house had become a home to us by the end of the summer
of 1977. I had ploughed the large garden of over an acre, imported
soil to make some of it more fertile and sown trees and shrubs. We
walked along the canal with Aoife and Mary, places where old boats
in various states of disrepair were worked on by owners at weekends.
But there was no Mrs Daly to share her words of wisdom or point
out the herbs along the canal banks.

Strangely, it was only then I came to realise that maps are about
boundaries. The solid lines of the map depict the boundaries
between fields, between houses, between townlands, between
baronies and between counties. The dashed lines that represent
imaginary boundaries of townlands and baronies running through
open fields render them solid in the mind. Fields, although not
named on maps, are named instinctively by the viewer: the upper
field, the haggard field, the marshy field, the well field, the tank field,
setting one apart from the other, endowing each with a character
and personality of its own.

Maps depict the boundaries between people's possessions. Yet
nature rarely respects man-made boundaries. A river in flood or
the sea in its rage can, in an instant, wash them away. Even a gentle
meandering stream like the Rye can eat away at man's efforts.

Crows cawing from rookeries, blackbirds singing in the
bushes send out their tunes across the landscape, unaware of
man's territorial obsessions. The robin flits unhindered from one
townland to the next. The fox, freely wandering the land, slips
through a run it has fashioned in the hedgerow of a townland
boundary. But man divided the land into parcels and the Ordnance
Survey recorded the divisions for posterity, making them at times a
source of conflict.

Perhaps boundaries were the reason I was attracted to a

career in mapping. The real and imaginary boundaries that had fascinated me as a child in the streets of Waterford had left their imprint: boundaries between rich and poor parts, invisible electoral boundaries, drainpipes and railings. The real was solid; the imaginary became real. And there were the nondescript 1930s streets around Griffith Place where my grandparents lived later in life, the endless terraces of anonymous, grey, two-storey houses fronted by small gardens and low iron railings.

Yet despite the anonymity of the terraces, each of the houses attempted somehow to state its own separateness. The paintwork, the garden, the hedges, the curtains hinted at individuality. Although most were minor variations on the others, there were some, though few, notable extremes. At one end of the spectrum were houses with regularly painted glossy windows, doors and iron railings, curtains that were fresh and neat, gardens that were cultivated and tidy, coloured with flowers and shrubs as the seasons permitted, displaying a pride in themselves. At the other extreme there were houses displaying tired and uncared-for doors and windows, paint long faded and now peeling, curtains framed uncaringly by those windows, gardens wild, unkempt or strewn with broken belongings, portraying the misery that had befallen those inside. Consciously or unconsciously these differences separated one family from the next and created barriers that could not be crossed.

When I went to the country as a child I was fascinated by the way I could wander at will through fields and gaps in ditches and suddenly meet a solid hawthorn boundary that stopped me in my tracks. Search as I might, there was never a way through. The far side was another man's land. Ancient boundaries, marked by stone walls built in bramble and furze-covered land, over knocks and up mountainsides made no sense except to the stones that had been carefully laid side by side or one on top of one another. The stones held the secrets of the hands that had rooted them from fields and balanced them into their exact positions. In other places clay banks, often worn by animal hooves and weather, were substituted for stones. The boundaries, in their tired, worn state, seemed meaningless but I imagined that they must still hold the echoes of conversations of the men who sat resting from their toil on the soft brown clay.

During my first field journeys as a surveyor, I met men and families to whom the ownership of land and the boundaries within which it was contained meant more than anything on earth. The land was theirs and they would have no one encroach on the sacredness of their trust without their explicit consent. Farmers, clergy, and self-styled princes alike had this passion for territory. Here are three of the most memorable.

THE OLD MAN'S LAND

In 1975 I was looking forward to doing some survey work near the Metal Man in Tramore, County Waterford. I had an ulterior motive for this. I could revisit a scene of my childhood where we often hopped three times on one foot around a pillar on which the Metal Man stood and wished for dreams to come true. Three stone pillars, the Metal Man perched on top of the central one, stand on a headland at the edge of Tramore Bay. They were built in 1823 to warn ships to stay away from the dangerous cliffs and rocks below. We were told that the Metal Man called out over the ocean on stormy nights; 'Keep out, keep out, good ships from me, for I stand on the rock of misery.'

It seemed an ideal point for a trig pillar as there was a good line of sight along the coast west to Dungarvan Head and east to Brownstown Head on the other side of Tramore Bay. I was a little surprised to find the access blocked with barbed wire. Scraps of galvanised tin hung from the gates. The words 'Trespassers will be Prosecuted' were roughly painted on them.

I called to the owner, who lived opposite, and began to explain my business. His face hardened and he stopped me in mid-sentence. 'There's no one going in there,' he said abruptly. 'Are you one of them fellows from Dublin who was here before? I gave ye yer answer then. And that's it.' He sat beside an open fire in the dark kitchen of his farmhouse. The fire did not share its heat with the room and scarcely gave it light. I was not invited to take a seat.

'I was never here before,' I replied. 'You must be mixing me up with someone else.'

'Who are you then if you're not one of them from that Office of Public Works?'

He softened slightly once he heard I was not 'one of them.' He told me he had been tormented by people from OPW attempting to negotiate a right of way across his land to the Metal Man to make it into a tourist attraction. There was emotion in his voice as he spoke about his land and the land his family had owned. Parting with some of it would have been akin to losing a limb, an eye or an ear. Having hundreds of people trampling his beloved land each year was unthinkable. He had blocked the entrance a few years back because people were not just walking along the old access lane but throughout his property.

It took me two hours and much repetition to explain that we were updating the maps of the county and wanted only to record the changes since the previous survey at the end of the last century. I fetched a map from my car and let him walk through it.

'Those ditches near the coast are no longer there,' he said. 'They were washed away in the 1940s. And that rock, it was once connected to the mainland ...'

He let me enter his land but it was unsuitable for the purpose of my survey. I could see along the coastline but many inland observation points were hidden from sight. Still, I got to be at the Metal Man, hop around it three times and make a wish.

PRINCE OF THE SALTEE ISLANDS

Wexford was trickier to survey than the Munster counties. The land in the south-east of the county is relatively flat, something which made the selection of sites for trig pillars and lines of sight difficult. The Saltee Islands off the south coast of Wexford seemed to have some potential as the entire coastline from Carnsore Point to Hook Head was visible from there. I assumed that there would not be a problem building a pillar on an uninhabited island and asked the survey team to proceed with the job. I didn't expect to become involved in a dispute about the sovereignty of a piece of land.

As usual, the survey crew made arrangements with the boatmen at Kilmore Quay to take us to the island. It was summer, the sea was calm and the weather fine, yet for weeks the boatmen were reluctant to take us out. 'The tides are not right for landing. It's a tricky place,' was their excuse more often than not. We could not challenge them. We did not know the island.

Then, about a month into our efforts, a letter arrived unexpectedly at Phoenix Park. It was from a 'Prince' Michael who claimed ownership of the Saltee Islands. Moreover, he claimed that it was an independent sovereign state. He had his seat on the island. He wanted the Irish state to recognise his claim before he would allow us on to the island. We explained that our job was to survey the land, not to determine ownership or sovereignty. After a long correspondence he relented, without relinquishing his claim, and allowed us to establish a modest trig point. The tides did not seem to be a problem on our next visit to Kilmore Quay. The fishermen gladly took us to the island and its population of rabbits and sea birds. We measured the distances to the trig stations at Hook Head, Rosslare and Carnsore and returned the solitude of the island to the wild creatures and Prince Michael's seat.

The Curates' Guns

We were commissioned to carry out surveying work a good distance from Dublin, which included constructing a number of new trig pillars on mountain- and hill-tops. We had a difficult hike to one of them, a barren, rocky, heather-clad hill with a long, energy-sapping approach through boggy land. The summit was hidden from sight and the survey team laid out ranging poles to guide our approach. High hills like this were usually commonage, shared by a number of local farmers. Here, however, a farmer living near the foot of the hill claimed full ownership.

We had tough negotiations with him about a price for the job and he eventually agreed to take the ton-and-a-half of sand and cement and all the other bits to the top with his donkey. It was difficult and the ass became stuck in the boggy ground. Everything was unloaded but the animal still could not be budged.

'There's no hope. I'll have to shoot him,' said the farmer forlornly.

He returned to his house to fetch his shotgun. But to his amazement the donkey was waiting for him at the door when he re-emerged with the gun.

It took three days of skilful navigation with lighter loads to complete the job.

Hugh McAllister and his team set out in the early morning to build the pillar and were puzzled to find the carefully-placed ranging poles smashed and thrown on the ground. I had gone to a meeting with the county engineer and I met a very sombre Hugh on my return.

'You're in big trouble,' he said and waited for me to ask why.

'It was like a scene from *High Noon* up there. Three men with shotguns were waiting on top of the hill for us. Everything – sand, cement and water – was scattered all over the place.'

'Come on, Hugh,' I said. 'Tell me another one.'

'No jokin'. One of them was very angry and the others nearly as bad,' he replied in the same gloomy voice. 'I thought they were going to shoot. We tried to explain what happened but it did no good. The one in the middle became even more indignant and said we were trespassing on his family's land.'

'Where does this farmer live?' I enquired. 'We'd better go and talk to him.'

'That won't do,' replied Hugh with a smile. 'You have been summoned to meet and explain yourself to one of the Reverend Fathers in the hotel there in town at eight o'clock tonight.'

It was true. Two of the men were priests and one of the priests claimed to be a son of a farmer who owned the back of the hill, including the summit.

'Come off it,' I said. 'We didn't upset anyone, did we?'

I arrived at the crowded hotel before eight, more curious than apprehensive. The holiday crowd had overflowed from the bar into the small lobby and porch. I had forgotten to ask Hugh for a description of the priest but assumed he would be alone and in clerical garb. There was no sign of a man with a Roman collar or black suit; nor had he appeared by half past eight.

The next morning Hugh and I visited the house of the aggrieved farmer and his family. The family – father, mother, sons and daughters – gathered at the door.

I introduced myself and attempted to explain. 'We made enquiries and understood that the land belonged to a man on the far side of the hill. I had no idea it belonged to you.'

'Ye have big jobs above in Dublin and ye think ye can come down here and ride roughshod over people like us,' declared one of the daughters of the house, who acted as spokesperson. There was anger in her voice but, even more, there was dismay in the eyes of each member of the family that strangers could trample on the land sacred to them and their forebears.

Nothing could have been further from the truth. It took all our efforts to convince them of our genuine error. In the end I suspected there might have been a touch of envy because a neighbour had been paid to bring the materials to their summit. With that in mind, and because the materials had been scattered, I decided that additional sand was needed. A deal was done, money was exchanged and the priest's family brought additional materials to the top, even though the distance was considerably greater than from the other side. I never found out why the priest did not turn up at the hotel the previous night.

10

THE SHAPE OF THE NEW IRELAND

The six-inch series was synonymous with the Ordnance Survey and captured the imagination of map users. The more detailed twenty-five-inch map series never had the same status. Yet it was as important as the six-inch series, being necessary for the legal transfer of property. Its principal drawback was that it covered a far smaller area of countryside: the six-inch showed the same area as sixteen twenty-five-inch maps. Only when the map depicted the patchwork of small stone-walled fields of the west of Ireland did it look anywhere near complete. Otherwise it was just a large sheet of paper showing a few large fields.

This map series began its life in 1888 at the height of the agrarian reforms in Ireland. Landlords needed very detailed maps showing the areas of each field, no matter how small, to transfer ownership of land to tenant farmers. The six-inch map was not accurate enough. It would have been necessary for landlords to produce maps at their own expense had it not been for the magnanimity of the then Director of Ordnance Survey, Major General Charles Wilson. He had inordinate sympathy with the landowners and sought to reduce their financial burden by recommending a new national series at a scale of twenty-five inches to the mile. He was obviously an influential man as in 1886 he persuaded a royal commission on Irish land law to approve. Twelve hundred men produced the twenty thousand maps that covered Ireland between 1888 and 1913 but they suffered the same fate as the six-inch maps in the first half of the twentieth century, being neglected by successive Irish governments. It was not until the 1960s that maps of a few select counties were systematically updated.

This map series had been chosen as the official rural map of Ireland in 1965 at the expense of the six-inch map. It was to be revamped and updated. The six-inch map was abandoned, never to be updated again, although demand continued for the six-inch series well into the 1990s because new updated twenty-five-inch maps were not being made available fast enough.

The six-inch map continued to have one important function. The original townland boundaries marked out by Richard Griffith in the 1820s had legal status and were plotted on the first edition. They would remain the source of clarification in boundary disputes.

The decision to abandon the six-inch series was lamented, particularly by traditional map users who had come to love the texture and feel and layout of the detail. It was the map on which the traditions of surveying and cartography in Ireland were built. People would talk to me about the six-inch map once they knew I worked in Ordnance Survey: about how great it was, about the great men that made it. They marvelled at how it was possible to make such a map in the early part of the nineteenth century.

I liked the map. I had been seduced by the map's history and by the copper plates on which the maps were depicted. They drew me into the great tradition of Ordnance Survey. But by the late 1970s I was becoming impatient to move on; I wanted to record the changing Ireland as fast as those changes occurred.

Modernisation dictated that the old name, the twenty-five-inch map, be converted to a metric scale. It became known as the 1:2500 series or more popularly as the twenty-five hundred scale map. One metre represented twenty-five hundred metres on the ground. Townland boundaries, county boundaries, houses, roads, bridges were to be checked; things that had disappeared over the previous fifty years deleted and new features added; names of new roads added; fields resurveyed; the areas of changed fields recalculated and the streams that meandered through the countryside and that had altered their courses remapped. There were more than twenty thousand maps in all to be updated. It would be a slow process involving a walk through every field that appeared on each map. We estimated that it would take over twenty years to complete, a lifetime to my mind. Gerry Madden's estimate of the time still stood. I was already looking for faster ways.

At least the time scale was better than what had appeared in the Ordnance Survey annual report for 1935. Then it was estimated from the number of maps updated for that year that it would take one hundred and fifty-three years to complete the remapping of rural Ireland.

The new rural and urban maps of Ireland differed in many ways from those made by Thomas Colby. All new mapping was known as the National Grid Series because the island of Ireland was treated as a single entity. There was one set of coordinates for the entire country and one map could be laid down beside the other to form a single continuous map.

Thomas Colby's six-inch and twenty-five-inch maps had been produced by treating each county as a separate entity and mapping it independently. There were thirty-three independent sets of maps, each with its own coordinate system because Tipperary had been divided in two, North Riding and South Riding. The country was like a thirty-three piece jigsaw. But the pieces did not fit together. This is because every county lies on the sphere of the earth and is slightly curved. Its shape is marginally distorted when it is projected on to a flat sheet of paper. So now the pieces had to be adjusted to fit the new national grid system.

Old trig marks held the key to transforming the old into the new. The marks placed by Colby and his successors were uncovered and resurveyed, using the national grid coordinate system, in order to transform the old maps. These marks, red tiles, buried principally in fields, were part of the original triangulation scheme for the first six-inch maps. There were hundreds in each county. The bigger triangles connecting mountain summits had been gradually subdivided until they became small and compact. They were then used as reference points for mapping detail: field boundaries, houses, streams and roads.

To locate the marks, field surveyors became virtual archaeologists. Patience, a delicate touch, a spade and a trowel were the tools needed. Little red tiles with triangles inscribed on them buried in fields where ditches and fences might have disappeared were not easy to find. The original written description was useless if field boundaries had been removed. It was difficult not to disturb the marks during the dig. I was always amazed at the surveyors' air of excited expectancy, no matter how long they had been engaged in this work, even though this expectancy that did not always result in a find. But I was never at a dig at which I did not hear a excited shout – 'Found it! Found it!' when a layer of soil was lifted to reveal its red secret. Every member of the team would file past and pay homage to the little red tile in the brown hole.

The transformation from the old to the new laid bare what was known but rarely admitted – that there were errors in the old maps and in some of the old trig marks. Occasionally the locations of field boundaries shown on the old map did not correspond to their positions on the ground. A kink or twist in a clay bank might have been missed or a chain-length dropped in measuring from one side of a field to the other. Nobody, not even the succession of landowners, might have noticed these errors in the one hundred and fifty years of a map's existence.

The county boundaries were often problematic. They needed to be resurveyed to ensure uniformity because of the way Thomas Colby produced his county series maps. There were often other issues where the centre of a river or stream marked the county boundary. I found in many instances that small errors might have accumulated when a county was originally mapped and that the errors were eventually buried in the rivers. I had first-hand experience of this at the River Blackwater on the boundary between Waterford and Cork where we discovered a two-metre error in the width of the river on the old map. My first reaction was to doubt the accuracy of our own surveying and have it rechecked but the two metres did not disappear. There were other instances, particularly with the major rivers. But it has to be stated that for most part the old maps were accurate.

Engineers, architects and property owners welcomed the

new national grid maps. They no longer had to buy two maps if their land or business straddled the county boundary and they no longer had to pay the full price for large sheets of paper near county boundaries that contained only slivers of maps. Engineers could purchase the maps on film for the first time, something which allowed them to superimpose their design schemes and copy them more easily.

The new maps were functional and up-to-date, with sharp lines, bold text and strong black ink on white paper that was less substantial than the older paper. The style of the old maps – ornate text and symbols, fuzzy black lines with character on thick, off-white paper – was gone. The new style reflected the developing Ireland of the 1970s, an Ireland casting off the shackles of the past, an Ireland with a mission to get on with things. This Ireland had replaced old with new and paid no attention to the ornamentation of the past. New developments and housing estates encroached on nature's serenity without respect; buildings which gave a place its individual identity were destroyed and replaced by ugly modern structures; whitewashed thatched cottages were replaced by 'bungalow Ireland'; ancient field patterns overflowing with history and memory were pushed aside to make way for modern farming. There was no sentiment in the new maps, no acknowledgement that an ancient landscape and its memories and traditions were being destroyed. The maps were just a factual record of the present. And when the present changed it too would be wiped off them.

The new maps of Waterford, my childhood city, showed nothing of what the developers had scraped from the landscape, only the hurried expediency of the replacements. The Victorian railway station, majestic in its proportions, the city's distinctive gateway to Tramore, flattened so that in time some shapeless monstrosity might be built; the railway tracks where children excitedly screamed and laughed on puffing steam trains, gouged out to make way for roads. Fleming's Castle, which overlooked the city from Mount Misery, bulldozed, a glass box posing as a luxury hotel rising in its place. The distinctive Bishop Foy School on the Mall levelled: an undistinguished office block of glass, steel and concrete deemed to look better. The Adelphi and Imperial hotels, which had preserved the elegance of the Georgian age, replaced by modern concrete. There were no radical students, as there had been in Dublin, to protest at the destruction. There was no occupation of the buildings to protect them. Nobody seemed to care.

What was gone was part of my childhood. They were the images I had implanted into my mental map of Waterford. They were memories of the days and nights I had wandered and played in the city and dreamed of what it might have been like to live in Fleming's Castle, eat and sleep in the Adelphi Hotel, or wondered what it was like to attend a Protestant school like Bishop Foy's. The Waterford map of my mind was rich with the memories and dreams and the energy of those dreams was stored there too. The new map, factual in the extreme, acknowledged nothing of those memories.

The destruction of buildings spanning two hundred and fifty years gnawed at me but at the time I cast it from my mind. I was happy to be the architect of the new maps that showed the present and hid the past. My childhood memories, some of which I was not yet ready to deal with, were in the past. I identified too many memories with my relationship with my father, who died suddenly on one of the city's streets when I was thirteen: the shop from which he was taken to the hospital in the back of a van; his bicycle lying against the shop wall which I had to collect the following day; and his funeral procession through all the streets of memories where we had walked together. His knowledge of the city and its traditions had become mine as I grew up. Some of the memories had reawakened when I discovered the one-inch map, Number 168, lying in the corner of a yard in the Ordnance Survey in 1972 but I managed to prevent the emotion that find represented from emerging from my depths.

I was happy to join in the mood of the nation in the 1970s and let the new smother the old. I wanted to use the cold edge of my mathematical and computer skills to speed up the production of new maps. I was willing to let go the attachment I had to the maps and copper plates that had seduced me when I arrived at Ordnance Survey. I needed to put the shape of the new Ireland on paper.

11

SIGNALS FROM THE HEAVENS

The Americans came in 1975 with their Doppler System, an early form of Global Positioning Systems or GPS, as it would become commonly known. They spent five days on the Hill of Howth communicating with passing satellites.

The British Ordnance Survey came to help six months later. Malachy McVeigh and his survey team travelled Ireland with them and reoccupied twelve of the primary trig pillars set up by Muiris Walsh in 1962. Together they sat at each station, in a Land Rover or a tent, in sun, rain, wind, fog and storm, twenty-four hours a day, communicating with every satellite that passed overhead. The survey teams were burned by sun and wind; some smoked cigarettes to pass the time. Their tents were often wet and offered little shelter from the cold of night. A Calor gas ring boiled the kettle and the eggs. An hour asleep and then up to track a passing satellite. Ireland was being tied to Britain and both countries to Europe. The Irish triangulation network of 1962 was becoming part of the great European Triangulation Network, the common Europe.

It wasn't easy. The primitive GPS computers had to communicate from each trig pillar with three satellites at a time but the satellites' appearance over Ireland was periodic and it took a week at each station to communicate adequately. The perspective was global. The new tools of surveyors were black boxes communicating with the heavens, juggling numbers and pushing out coordinates.

The weather didn't matter, as signals travelled daily to and from space, through cloud, mist and the troposphere, stratosphere and ionosphere. Disturbances in the ionosphere attempted to disrupt the signals but mathematics corrected that. The geometry of Ireland was now in the heavens; the triangles of the hills of the landscape were not needed. A ground point and two satellites formed the triangle and the coordinates were calculated.

Ordnance Survey men communicated with the satellites, waited for them to peep above the western horizon and disappear through the eastern doorway. Sometimes there were none, only a long wait by day or night. The men stayed at the station; there was no time to go to the village. In the middle of the night at Carrigfada in County Cork they listened to iron gates being opened, footsteps without voices passing from the road to the woods beneath the hill. Sharp cracks and more cracks; long bursts, short bursts, single shots. Footsteps disappearing into the night: training in the Republic of Ireland for IRA activities in Northern Ireland.

Next morning, the farmer asked if they had gone to the village during the night. 'One of my gates was opened,' he said.

They replied, 'No. We saw nothing.'

'Strange,' he replied, 'I could swear I heard something.'

Nobody ever saw the IRA.

Speeding satellites, number one up, number three up: no lyrical names like Carrigfada or Carrigaderg; just numbers, devoid of character. They had no long history to look forward to, only the fall some day towards earth and burning in the atmosphere. A replacement would be on its way.

Other satellites took photos of the country. These consisted mostly of cloud but the optimists argued that the satellites could photograph the entire island of Ireland in minutes; a blue sky was all that was needed. They forgot that the satellites could not wait for Ireland's clouds to clear. Academics pieced fragments of Ireland together and made a picture. The Americans made satellite maps. They thought that this was the future of mapping and that the surveyor did not need to walk and measure the land any more. But their satellite maps were general and could not match the detail shown on the six-inch or the twenty-five-inch maps. When enlarged they broke up into meaningless pixels.

I was excited by satellite maps but even more so by the high-altitude aerial photography of Limerick that Ordnance Survey had obtained in 1973, and which the Geological Survey commissioned for the entire country in 1975. This photography had possibilities far greater than any satellite image. Maps could be made and updated from it and information on the photographs could provide enhanced details.

I remember the day I first saw them laid out in the air-survey office. Images that up to now could only be fleetingly seen from a speeding aircraft had become a new historic record of the landscape. Photographs of Limerick from fifteen thousand feet – clear monochrome pictures, twenty-three centimetres square – showed far more than a map could ever depict: vegetation, ploughed fields, pastures, marshy lands, bogs, hedges and walls, old and new houses.

We saw the regular pattern of new houses on half-acre plots not yet matured by trees and hedges; the random shapes of the boundaries around old houses and farms whose characters were displayed by the layout of buildings, trees and hedges; manor houses with parklands of trees and pasture; others suffering from neglect. Old graveyards, new graveyards – we distinguished them all from the photographs.

Ring forts that had not been mapped by Thomas Colby's men. What could not be seen when walking the ground emerged from the aerial photographs. Our ancient heritage was re-emerging.

'Look, look, Captain Kirwan,' said John Danaher, the manager of the air-survey office, as he rushed into my office one day brandishing two aerial photographs. He pointed to a dark, broken ring. 'It's a ring fort. And look, there are more,' he said, indicating large discontinuous rings a deeper shade of grey than the surrounding grasslands, which were clearly visible. Ring forts could not be identified by walking the land because they had been levelled or weathered away over the ages. But now they had come back to life.

The photographs were so striking that my mind didn't take in that they were monochrome, superimposing various shades of greens and browns and blues that made what I was looking at real. Aerial photography was the new way forward. Photographs and maps would complement each other from now on. The map was the linear representation of the land. The photograph was a pictorial snapshot of the landscape. There was only one problem. Ireland usually had no more than twelve days each year with clear blue skies and it could take at least five years to photograph the country completely.

There were plans to produce a new contoured map of Ireland using aerial photography. Ireland would be contoured for the first time since Thomas Colby's men drew contours on the six-inch map. The new contours would be at closer intervals than the original ones. Engineers wanted a detailed contoured map. It would be good for planning new roads and drainage schemes and improving the country's infrastructure.

I could imagine other things. I could envisage the ancient landscape of Ireland rising from a flat sheet of paper endowed with these contours. The Hills of Tara and Uisneach, the ancient seats of high kings, could be brought to life. Newgrange and Lough Crewe, where our Neolithic ancestors sought the divine in summer and winter solstices, would jump out from the paper. The soft undulating drumlins of Leitrim and Monaghan would be there. And in my own County Waterford, the low rocky outcrops of Sugarloaf, Ballyscanlan and Smoor, where I lived in childhood imaginings, would be discernible.

One map, part of County Limerick, was published. But no further maps were published at this scale because the time and cost of production was far greater than anticipated. Ireland would have to wait until the late 1980s to have detailed contours drawn on a map.

I foresaw great possibilities for Ordnance Survey with developments in GPS and aerial photography. I was excited when I considered all the possibilities and became impatient. If someone questioned my new enthusiasm, I would ask them why anyone would want to dwell on memories of the past – they were for bygone generations sitting by the fireside. The future was all that mattered.

Why would I want to spend my days in an office filled with people sticking placenames printed on plastic strips on to maps? Or have another office full of people measuring the areas of fields, every day, every week, every year like their predecessors did since 1824: two million fields, a lifetime's work? Or why would I want to let the must of the past that filled Mountjoy House influence the way we made maps? And why, oh why, would I want to continue to make each component of the map separately when in the future we might make the map in one go?

Why should I climb the hills and mountains, battle with the mountain streams and rock faces, bog holes and squelching rushes to take measurements when I could sit in a Land Rover by the side of the road and let GPS receivers pull the signals from satellites? The time was near when more satellites would be in orbit and reliable positions and coordinates would be obtained in minutes. I could sit and watch a white disc on a tripod send invisible signals into the sky and use the triangles of the heavens to calculate the coordinates. After all, I no longer needed a line of sight between pillars. I could pick any point and send signals to the satellites. These went faster than the speed of sound and the Land Rover only had to move to the next station and begin the communication all over again.

And why should I slog through fields and briar ditches and stumble through long grass and drag a measuring chain from one end of a townland to another when an aerial photograph in a machine could do the same thing ten times faster?

Whisperings about the new signals to the heavens might have taken place between grass and leaves, honeysuckle and hawthorn, blackthorn and ash, chestnut tree and hazel. But I had no interest any more in watching and listening. A butterfly landing on the white disc of a GPS would no longer provoke my curiosity. I would not have the time to wonder if it understood the signals to the skies.

I could beat the forces of nature and would not, like Thomas Drummond on Slieve Snaght, have to hide from the gale and rain and wait patiently for the fog to clear to shine his light. Our signal would penetrate fog and mist.

My vision was simply that a map be made with as few hands as possible, from satellite signal and photograph to paper.

The mapping agencies in England and Canada had got rid of the pen and the scribing tool. It was called digitising. A small disc with a crosshairs sent invisible signals to a computer, which registered the boundaries of the map. They were called coordinates. Fields and houses and rivers became sets of lines. The computer had to be told what the line represented and did not accept hidden memories of the landscape. It was my hope that Ordnance Survey in Ireland would soon follow down this path.

12

New Challenges

Since the mid-1970s, Muiris and I had watched and waited for the world of computers to become affordable for us. We had looked, listened, learned, analysed and met those who had begun to computerise their maps. We heard the outlandish claims of fledgling companies who boasted that they could without difficulty automate map-making at Ordnance Survey. Their glossy brochures looked convincing, their presentations at conferences were compelling and their verbal claims extravagant. We listened for the more important signs: what was not said or avoided. People are not always willing to disclose difficulties.

Ordnance Survey of Great Britain and other bigger mapping agencies had been active in computer mapping since the early 1970s but we hadn't been convinced by what we saw. Most agencies were developing their own systems. They employed in-house programmers and computer experts and used mainframe computers. The early versions of their software had serious limitations. We could not justify that kind of investment. Besides, we wanted to retain map-making as our core skill. We wanted tools to do the job. We did not want to have to invent or make them.

Throughout those years my vision for Ordnance Survey became more ambitious. People spoke of the emerging information society and paperless offices and this sent a buzz through me. I could see the Ordnance Survey of the future filled with computer screens, whizzing magnetic tapes and automatic drafting machines. I was sure the cartographers we were now recruiting would want to be part of this new world and bring the Ordnance Survey legacy of enthusiasm to it. But it was impossible as long as computers were not affordable to relatively small organisations like ours.

The past and all that had seduced me – the copper plates, the printing skills and the printing stones and the maps that came from them – began to fade. My energies went into looking to the future as I knew the legacy of the past could not sustain Ordnance Survey for long more.

Finally, in 1979, we had the opportunity to introduce computer mapping to Ordnance Survey. The looking, listening and hoping were over and my dream was about to become reality.

The technologies associated with computer mapping had advanced significantly. European and American companies were developing new systems and software. There was competition and the prices reflected that. We knew that the Irish Ordnance Survey would be a good catch for some company, a good reference site for future business. A national mapping agency had status. Ordnance Survey could take advantage of this to get the best possible price and influence developments to suit our mapping needs.

Coincidentally, about then I moved up the management line. I took over as deputy to Muiris Walsh, even though I was still commanding officer of the survey company. My promotion had been accelerated by the sudden death of Gerry Madden, the departure of Eoghan O'Regan and then the retirement of Muiris's deputy, Tom Casley. I was only thirty.

I moved to a new office beside the busts of Thomas Colby and Thomas Larcom in Mountjoy House. Their presence outside my door was an inspiration. I was sure that they too would have undertaken the journey we were embarking on. I needed their presence to sustain me, as my new office was stark and uninspiring, with a cracked, uneven ceiling stained by water from a leaking tap overhead and walls covered in cold, white floral paper. A frosted glass door and window made it seem like the waiting room of a railway station and a hollow ringing echoed there whenever the door was opened or closed. The mahogany desk and conference table were lost in this big room. Nevertheless, I was to spend fifteen years there and implement the renewal of Ordnance Survey with Muiris Walsh.

During the previous years Muiris and I had established an easy and informal working relationship. We moved freely between each other's offices, dropping in on each other when there was something to discuss, but mostly we conducted business over coffee in his office each morning.

He put his head around my door late on a sunny Friday afternoon in September 1979 and invited me to come to his office. There was something different in his look, a look I had come to recognise whenever he came up with a bright idea. I knew this was not an ordinary meeting. White china cups and saucers were on the coffee table as usual and the silver pot gurgled as it did every day. He had poured two cups of thick, bitter coffee, topped by an oily film from the day's brewing.

'Well, how about digital mapping?' he said after we had discussed our weekend plans. He was going to play golf at Royal Dublin and sail his yacht in Howth Harbour. I was going to spend my time landscaping my one-acre garden in Hazelhatch. 'Do you think it's time to take the plunge?'

I did not have to think too hard as my mind was made up before the question was asked. I knew we had the resources in our budget. This was the opportunity I had been waiting for: the chance to completely modernise Ordnance Survey and all the maps of the country.

Muiris responded cautiously to my enthusiasm and refused to accept my immediate answer.

'Sleep on it. It is a huge decision. Your life and the Ordnance Survey will never be the same again if we go this way.'

There was no compulsion to follow this path. This was a personal decision for both of us: the ambition of two men to bring Irish mapping into the late twentieth century. If one of us had said no the status quo would have remained while we worked there. Computerisation would not have taken place until some other adventurous spirit came along or outside circumstances forced Ordnance Survey to change.

There was a choice for me. I could sit at my desk until I was sixty-five and ready to draw a pension. I could tinker with the old maps and blame the Department of Finance for not funding Ordnance Survey properly. Or I could accept this challenge and use my talents to the utmost to modernise the maps.

My nature and upbringing made the choice. Family circumstances had toughened me. My mother had emerged from the shattering blow of my father's death and gone back to work to educate my sisters and me. My grandparents had lost their home and their tailoring and grocery businesses in Michael Street in Waterford. They rose from the ashes of that loss and built a thriving printing business. How could I sit back and do nothing?

In my excitement, I did not consider how my decision might affect my life or my family. I looked on what I was doing as part of the greater modernisation of Ireland, something that was rapidly gaining pace and could only be good for my children and for future generations.

That weekend I was preoccupied with the cultivation of our triangular garden at Hazelhatch and with how we could make the house and the evolving landscape look as if they had been there for generations. The lawns had been sown and we were limited in what else we could do because of the poor soil.

It was tree-planting time. We wanted to break the monotony of a bare lawn with lots of trees that would draw us into the garden and provide a welcoming feel. We wanted colour in the summer and form in the winter. I planted oak, birch, beech, sallies and weeping willows. I set out copper beech and purple maple and dwarf acers that would make the garden more interesting. A weeping cedar, a weeping beech and a witch hazel would stand out against the others. I wondered how I might divert the stream, the home of the townland boundary, into the garden and let it weave its way through the trees. And the girls might like a bridge or two over it. I planted fast-growing leylandii to shelter us from the road, not realising that they would plague me with rapid growth.

With all the work we didn't have time that weekend to drive to the shores of Lough Ennell, near Mullingar, where we often picnicked and the children swam. Our family had grown and our first son, Richard, was in the pram. Instead, we walked the towpath of the Grand Canal near our home. This was the boundary between Commons, our townland, and Hazelhatch. They were connected by one of Lord Cloncurry's humped-back bridges. A man wearing a sailor's cap worked on a rusting barge near the bridge. We wondered how many years ago the barge had plied the canal, where it had travelled and what its cargo had been. A second man was restoring a timber hull that might have been better left to rot. Another boat lay badly listing in bullrushes, forgotten by its owners. Further on, the swans glided towards the canal bank, hoping that Aoife or Mary would throw them crusts of bread. And all along the bank broad-leaved plants grew. Aoife named them winter heliotropes, so that is what they were to us.

'Yes,' I said on Monday morning when Muiris and I met for a cup of coffee in his office. The coffee was strong and fresh; it had not yet had time to stew or form its oily scum.

'I thought that's what you would say,' Muiris replied. But he cautioned me again, repeating that it would not be easy and that it would take up a lot of my personal time.

The fate of Ordnance Survey was sealed by a day on either side of a sunny weekend spent by one man gardening and dreaming at the Grand Canal and by another yachting and golfing near Howth. I sealed my destiny for twenty-five years and Muiris his for fifteen. Thomas Colby would have his maps renewed. I wondered if he had listened from his pedestal, not far from Muiris's door, or if his ghost listened in the empty chair at the coffee table. Drummond and Colby might have swapped ideas about the new equipment they could design to assist the process. Larcom might have hoped that we would include a memoir in this new adventure. They all knew what it meant to introduce new methods to Ordnance Survey and I am sure they would have approved.

Six days later Muiris and I met the computer experts from the Department of Finance. We had a simple story and articulated a single ambition. We wanted to make the map-making process more efficient in order to produce more maps. Computerising the process would enable us to do that. We left one hour later having got their approval for our plans.

We had taken our first steps in bringing Ordnance Survey into the information age, leaving behind the age of the industrial revolution that Thomas Colby had introduced to map-making. I called our new venture the quiet revolution, a period in which we would question everything and our world would gradually change for ever. Quiet because we did not do it in a blaze of publicity. We would be working at the edges of technology with all the risks that entailed and I did not want to draw too much attention to what we were doing.

We had a different emphasis from that of Thomas Colby. He and Larcom envisaged the original mapping of Ireland in the nineteenth century as part of a greater project. Their vision was to record the country's geology, social and economic conditions as well as its history and other statistical information. Larcom attempted to achieve that in his memoir project. Theirs was to be a record of the past and present, a snapshot in time.

Our revolution had moved on from there. We wanted to record the new present on the maps and we wanted to anticipate what sort of maps users would require in the future. We were looking to the needs of the changing Ireland, of businesses and consumers who demanded more and wanted it quickly. We set out to do something different with the maps. We decided to add intelligence to them so that the emergency services would be able to do route planning and insurance companies could see at the touch of a button where flooding was likely to occur.

There were personal and organisational risks. A major transformation into computer mapping might not work, particularly as the cartographers had virtually no background or experience with computers. Change might take too long to implement or, worse still, distract from the major task of updating the maps we already had. The systems might not work or the companies we dealt with might fail. These companies were operating at the frontiers of technology and a wrong decision could drive them out of business.

But I had an unshakable belief, which Muiris shared, that

we had no option but to travel this journey and that this was the opportunity for real renewal. I saw the top of the mountain and for me it was a straight run to the peak. There was a map waiting for me on the summit at least as good as the map of my childhood dreams. It was the new map Number 168 of Waterford, produced and drawn completely by the computer.

13

In Search of Perfection

The age of the mini-computer had arrived in Ireland and the dominance of computer giants like IBM was being challenged. The mini-computer had been first unveiled by an American company, Digital Equipment Corporation, or Digital as it was later known, in 1963. Its evolution was slow and only now, in the late 1970s was it becoming suitable for the complications of computer graphics and map-making. Other companies were developing these mini-computers, which were becoming as powerful as small mainframes and taking up only a fraction of their footprint. They were affordable for organisations like Ordnance Survey.

We were faced with a new, vibrant culture of computer companies developing mapping software. They were looking for new opportunities, quick profits and cash to invest in the future. The mapping world held a clear promise for many start-up companies with no history, no tradition and in many cases no loyalty to anyone but themselves. They had emerged from the dreams of one or two people. Some were spawned from bright ideas developed in university campus companies by sharp, progressive people, not frightened to go for broke and see where dreams would lead. Instant fame and fortune or a lucrative buyout by a bigger company beckoned. Failure was possible but these dreamers had the confidence to build on what was no more than a learning experience. They all had one thing in common: optimism that their company could beat the world and become the leader in their speciality.

The culture of the new companies was alien to most people in the mapping industry. Old-established firms like Zeiss in Germany and Wild in Switzerland had for generations supplied precision-engineered optical instruments to national mapping agencies. They manufactured theodolites, air-survey machines and aerial-survey cameras. They were as conservative as the people who worked there. Loyalty was their hallmark. Theirs had been a stable world where change was slow and barely noticed. Often the salesmen seemed more interested in displaying the quality of the instruments than effecting a sale. Their world was not unlike the Ordnance Survey I entered in the early 1970s.

We looked for companies that could help us. We did not have to search too long because these companies could smell a prospect in the wind. They invited us to visit: the good, the mediocre and the hopeless. A Sunday-night flight in coach class, Monday and Tuesday all day in air-conditioned offices, late Tuesday night a flight to Dublin, on Wednesday morning the office in Mountjoy House: this became our regular schedule as Muiris and I made endless visits to computer-mapping companies in Norway, England, Scotland, Holland and Germany. Mostly we travelled together. We needed two brains to assimilate all that was being promised us. We had to decipher what was real, what was aspiration, what was fairy tale. Technicians and computer programmers spent days before our visits conjuring up solutions that could never be practical.

We gave the companies maps of parts of Dublin around O'Connell Street, my own Nicholastown in County Kildare and the tiny fields of west Kerry. We expected to see the computer reproduce them exactly but the early reproductions were an embarrassment and bore little resemblance to the originals. The maps were just lines poorly drawn on flimsy computer paper. Houses lost their characteristic rectangular shape. It was almost impossible to differentiate between roads and streams: both were just two parallel lines. We were amazed that they could not have the name of a river follow its curves. The richness of hand-drawn cartography and symbology was absent. Townland boundaries were unrecognisable: the series of dots that traditionally described a boundary had become another line. What worried me was that the vendors were proud of their achievements. They thought they were brilliant and were hurt when we told them what we thought.

The fact was that most people in the computer companies had no background in surveying or mapping and little appreciation of the art of map-making or of the conservative nature of cartographers. They were salesmen first and foremost, whose job it was to sell computers and software and make a profit for their shareholders. Because of this, they were a convincing bunch. The sales teams were slick and professional. They wore shiny grey suits, starched shirts and flawless ties. Their black shoes were polished and so were they. They promised to do all we requested, to rectify the 'little problems' we had noticed on the first maps. They would say they had 'misunderstood' what we wanted when something was not to our liking or did not meet the high standards of cartographers. It was hard to get under the polished veneer of many of the companies and get to a real understanding of what they could do.

A few British and German companies were arrogant and told us what we wanted. They knew best. Worst of all, some British vendors assumed that the Ordnance Survey in 'Éire' was an extension of the Ordnance Survey of Great Britain and that we would follow exactly what they were doing.

Still, it was good that so many companies were interested in our business. They learned from us but most of all we learned from them. They came up with new ideas to make the computerisation of maps more efficient. We played the ideas of one company against the next so that the system we eventually purchased contained all the good ideas.

In the end there were only two contenders, Ferranti in Scotland and Kongsberg, a Norwegian company, both long-established, with experience in the automatic drawing-table business. There was little between them. Both employed people with mapping backgrounds and both developed a reasonable understanding of Irish maps. This was a comfort to us.

They would do anything to win the contract. Jim Williams of Ferranti, one of the few in the industry with a mapping background, turned up in Malmö to hear me speak about our new mapping programmes at an international conference. I was given the worst possible time for my talk, the last session on a sunny Saturday afternoon. Seven people turned up in a five-hundred seat auditorium: the chairman, three speakers and an audience of three. By the time I took the podium the audience had shrunk to two. Jim was one of them, sitting near the top so that I could easily recognise him. I could not have missed him for he had a habit of constantly twirling a lock of his thinning grey hair around his index finger. There were loud snores five minutes into my talk, a real confidence-booster for my first international speech. It was Jim and he did not come to until I was almost finished. He congratulated me afterwards, saying my talk was most interesting.

Muiris and I arranged a final round of clarification meetings to tie up loose ends before we selected the winning contractor. We could not travel as there was a strike at Aer Lingus but it was critical that we place an order before the end of the financial year or our funding would revert to the Department of Finance. Ferranti came to the rescue; their executive jet happened to be in Ireland and it took us to Edinburgh.

Unfortunately the pilot descended into Edinburgh airport too quickly, which aggravated Muiris's sinuses. He tried to listen to presentations but could hear nothing but words spinning around the room, bouncing off the walls and bombarding his throbbing head. To make matters worse we had also arranged to visit the Kongsberg offices in Maidenhead and had to take the train from Edinburgh to London. They had agreed to fly us back to Dublin if the strike was still on. A Ferranti salesman joined us on the train and overwhelmed us with more technical information, making Muiris's headache worse.

Next day we listened to endless presentations that did nothing to stimulate Muiris, whose headache hadn't gone away. He left it to me to probe the Kongsberg people about the issues they preferred to avoid. Kongsberg won out. Their suits were less shiny, their talk more open, their proposals and demonstrations more practical and a closer match for our hopes. They had allowed us access to their

Thomas Colby (1784-1852), Director of the Ordnance Survey.

Greenville: extract from the Kilkenny six-inch Sheet 43, 1843.
(With permission Ordnance Survey Ireland)

Cappoquin: extract from the Waterford twenty-five-inch Sheet 21-11, 1906.
(With permission Ordnance Survey Ireland)

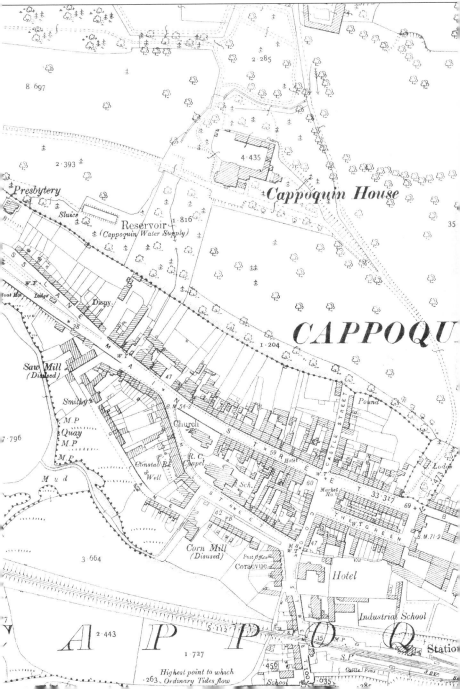

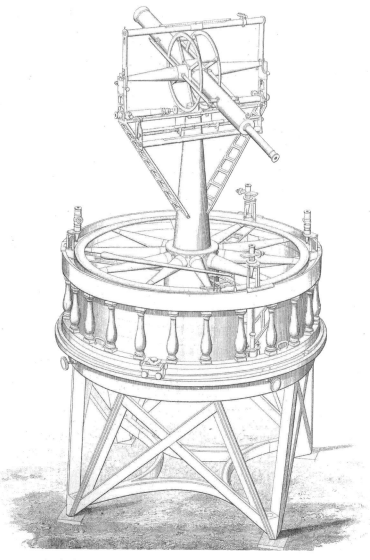

Ramsden's three-foot theodolite, used in the original triangulation.

Watch Tower, Waterford.

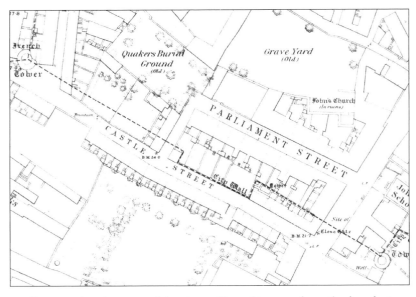

Extract, showing part of the city walls and towers, from the five-foot Town Plan, Waterford, Sheet 9-79, 1872. (With permission Ordnance Survey Ireland)

Extract from the five-foot Town Plan, Westport, Sheet 88-23, 1894.
(Reproduced with the permission of the Board of Trinity College Dublin)

Extract from the five-foot Town Plan, Killarney, Sheet 66-48, 1885.
(Reproduced with the permission of the Board of Trinity College Dublin)

Part of the urban 1:1000 map, Waterford 5701-09, 2009.
(With permission Ordnance Survey Ireland)

technical people, which enabled us to assess the strength of their proposals.

The introduction and improvement of computer mapping systems in the 1980s became a blur. Every year there were ongoing software problems to solve and new, enhanced, computer equipment to install. I became wedded to the new ways, thought about them in the car and garden, dreamed up new concepts. There was a mixture of excitement and frustration in implementing the systems and in the birthing of the new maps. They had basic lines with little of the ornamentation of the six-inch map but they were functional, they showed everything the engineers wanted, and, most important of all, they were up-to-date. All the new housing estates and roads were depicted on these new maps that had been revised by the field surveyors. This was our saving grace, which camouflaged the loss of the ornamentation and the imperfections of the computer-drawn maps.

In truth, we were unhappy with the development of the systems. The Kongsberg people had been over-optimistic about the capabilities of computers and drawing machines. We had a very simple requirement, or so we believed: to reproduce the Ordnance Survey paper maps faster by computer. Our specification had been too simple to quarrel with, yet there was endless argument over how it might be met. Kongsberg engaged the best brains from Norwegian universities to write new software and gave us an additional mini-computer to boost output.

The cartographers took to the new systems as if they were an extension of traditional drawing methods. Within three years almost everyone, young and old, had been retrained. I was surprised at how readily they adapted to the new technology. Tom Shanahan, a long-time traditional cartographer and self-taught motor mechanic operated the drawing machine. He treated it as if it were a car. The ticker tape that controlled it was like a timing belt to him. 'The belt broke last night and I had to strip her down and put on a new one this morning,' he would tell me in his verbal report. Or he would say: 'She needs her thousand-mile service. The nib was a bit off last night.'

Drawing techniques perfected for more than a hundred and fifty years were swept away over a three-year period. I remember saying to John Danaher that we would be in trouble if we had to revert to the old methods. Nobody would want to do it and after a few years nobody could. There was no cry to retain specialist units that could preserve the traditions of the past.

I became both a driver of computer development and a slave to it. I lived by its rules of incessant development. I loved it. It was my new god: it would solve all problems and make everything possible. I heard the gospel and I preached it. I listened to like-minded people at conferences in Ireland, England, Norway, Germany and Holland. I preached the same gospel at those conferences. I lived with the problems and with the successes. The work infiltrated the landscape of my garden. My mind did not stop to talk to nature, to listen to the rocks I had so carefully placed, to hear what they were saying.

Flying to Norway became part of my life. The Grand Hotel in Kongsberg: marinated herrings, tepid hard-boiled eggs, slivers of cheese and dark, coarse rye bread for breakfast, on every visit, year in year out. We pushed Kongsberg to the limits, never satisfied with what was on offer, always looking for the latest developments in computer mapping. I wanted perfection when it was not yet developed. The industry was in its infancy but every year it could do more; every year it could add new intelligence to the map.

The Ordnance Survey and the Norwegians engaged in brain-storming sessions. We wanted to find new customers and new uses for digital maps and talked of giving the map more intelligence as if it might some day speak to us. We envisaged supermarket chains bringing the map to a screen at the touch of a button, seeing where the new housing estates were built and planning for new shops and product promotions without ever visiting the area. Insurance companies in far-off lands could look at a screen, check if a house was too near a river and refuse cover. There would be no need for human contact, as all the information needed to make instant decisions could be pulled from a computer screen. I never stopped to consider whether I wished to contribute to this new society. There was a tide of modernisation, a consumer society, and I would play my part in it.

During those first years, we engaged in tough, draining negotiations, especially when the installation of the first computers fell behind. The Norwegians engaged wave after wave of technical experts to camouflage any problem and it was difficult for Muiris and me to absorb all that was being presented to us.

We visited Kongsberg at the beginning of our second year of the implementation. It was January and twenty-eight degrees below freezing but it was twenty-five degrees in the office and we were in shirtsleeves. For the whole of the second day we discussed tricky parts of the contract that had yet to be implemented. I was wilting. Muiris was tired and decided to add a little spice to the meeting. 'When will you have the operating manuals in Irish ready?' he said. He looked at me and I said we needed them immediately. There was silence. The presenters looked at the project manager. He looked blankly at the table and said he had not seen it in the specification. Muiris told him he had better reread the documents and he left the office.

Muiris smiled. We poured black coffee into white cups. It was strong and refreshing and pumped the adrenalin again. We put on our coats and stretched our legs on the cold veranda. It was ten in the morning and barely light. The sun would not peep above the hill until eleven and then it would creep up the brow of the hill without showing its full sphere and disappear by two o'clock. The river below was frozen; snow lay on the ice. Fallen trees had lodged by a small island. Agitated waters broke from under the ice a hundred metres away and thundered over the rapids near the town bridge, the spray held still in the air. A lone car passed over the bridge and a pedestrian wrapped in winter clothes walked to the other side. Dim lamps shone in the windows of wooden houses at the far side of the river. It was a sleepy town at the best of times.

Our coffee steamed in the freezing air. The project manager returned after an hour looking bemused. 'I could not find the section on Irish,' he said. That night they brought us through the snow up narrow winding roads to a mountain-top restaurant for dinner.

We had set out with a simple concept: computerise the maps,

make them easier to produce and get Ireland remapped faster. The maps were designed to be functional and give nothing more than up-to-date information to the users: those produced in the 1980s were just that. They were never meant to be like the six-inch maps which could evoke memories of the culture of a country and show its soul at a glance. But they did mimic the culture of the New Ireland, a society where everyone wanted to have the same as their neighbour. The new urban maps showed standard rectangular houses in uniform gardens; housing estates of similar proportions designed with little imagination, with the aim of squeezing as many rectangular boxes in as possible; standard footpaths; standard roads and lamp posts set out at regular intervals. Every house was numbered and the numbers made them anonymous. Even rocks along the coastline became standardised shapes.

14

Choppy Waters

The early 1980s were times of great expectations, successes and frustrations, of friendly meetings and difficult meetings. It was the dawn of a new era for Ireland and for Ordnance Survey. The cartographers expected improved conditions because of all that was happening and wanted to be compensated for the introduction of new technology, new skill sets and a total restructuring of Ordnance Survey which would give them better promotion opportunities. They felt hard-done-by as they had been overlooked when the conditions of cartographers in the Land Registry were improved.

Theirs was a reasonable expectation and Muiris was sympathetic but a major restructuring was not in his gift. It would have to be sanctioned by the Departments of Finance and the Public Service and they were in no mood to be generous. They had introduced an embargo on recruitment and promotions in 1981. The civil service had grown rapidly during the late 1970s and now the government was imposing cutbacks in the face of an economic downturn. The Department of Finance was adamant that the embargo be strictly adhered to, no matter what case we made about improvements in output, and demanded big cuts in our staff numbers, insisting that we transfer some cartographers to another department. In the end we gave up three cartographers but the real attrition started only when people began to retire or move to other jobs.

Muiris was faced with a problem that had the potential to kill the computer-mapping project and halt the resurgence of Ordnance Survey but he was skilled in the politics of persuasion.

He had the temperament for it and he believed in the justice of the cartographers' demands. I was only an apprentice in the negotiations that followed. He reasoned with the unions. He sat with them in his wood-panelled office and gave them coffee in china cups. It was not easy, for he could promise nothing and they became impatient at his apparent stonewalling.

Tom, a full-time union official, led the staff team and adopted an aggressive approach that included personal attacks on Muiris. Muiris ignored these and continued to reason, hoping to bring this man around, but this approach did not work and the exchanges were becoming nastier. At the beginning of the next meeting he set out negotiating parameters. 'Tom,' he said. 'Let's get the personal insults over first and get on with the real meeting.'

Muiris emphasised the justice of union demands to the Department of the Public Service but officials there did not want to know. 'There's an embargo on recruitment and promotions,' was their stock answer. They would not meet with the unions, as they said it was a local dispute. Through late evenings in the department's air-conditioned offices, Muiris repeated over and over, like a mantra, the justice of the demands of his staff. I chipped in with my arguments on increased output. The department officials still said no, no, no. There were threats of a strike. Already there was a work-to-rule in some offices and cartographers refused to use some equipment.

'It doesn't matter,' said Willie, of the Department of the Public Service. 'Play hardball. Give them nothing.' He repeated that they had nothing to give and that if they gave anything they would set a precedent for other departments. Muiris argued that there were special circumstances in Ordnance Survey and that the department's attitude would cause a strike. Willie said, 'So be it.'

The negotiations lasted months and months, through hot summer evenings and cold, wet, wintry ones. For me it was such a waste of time, when everyone knew that this thing would be resolved somehow. We thought we saw a chink in the department's armour at many of the meetings but when we tried to exploit it Willie sealed it and left us with his parting shots, 'Play as hard as you can. Concede nothing.' But despite this intransigent stance the department worried about the consequences of a strike and Willie was wont to shout as we closed the door behind us, 'Do anything but don't cause a strike.'

Muiris wore the department down and eventually secured a reorganisation but the unions made fresh demands as more sophisticated equipment was introduced. The 1980s became a difficult time for us. There were regular skirmishes and threats of equipment being blacked but we got the computer systems to work and increased the output of maps.

Our recruitment drive had ground to a halt with the embargo and we had only three hundred and fifty of the eight hundred and fifty people we needed. No additional cartographers would be recruited until 1997: seventeen years without recruitment. Nor did the military help with our staff problems. They needed more men for increased security duties in border regions and were also affected by recruitment restrictions. Muiris and I questioned the wisdom of the army presence. We could not afford casual absences for military duties. It seemed at times that we were being hit from every side, that we were the men in the middle with he unions, the department and the army around us, all with their own agendas, while our agenda was to get Ireland remapped.

In the face of the recruitment embargo and budgetary constraints the obvious thing would have been to reduce services, limit the development of computer mapping and set out a modified programme for renewal. Most of our customers would have under-stood. They suffered from the same embargoes. Instead, we ex-panded our activities. In 1979 I set up a new commercial con-tracting section. The national roads programme was in its infancy and Ordnance Survey had an opportunity to take an active part in drawing up the preparatory and planning maps. This would make us more dynamic and increase our revenues. The Department of Finance provided us with a 70 per cent subsidy and wanted it reduced. Commercial contract work was available in abundance in counties like Donegal, Kildare, Meath and Cork. We chased and won it. It put pressure on us to deliver on time while simultaneously having to meet our own targets for additional urban maps.

John Danaher came under pressure from everyone. He represented the Ordnance Survey to the county councils. He negotiated and managed commercial contracts and at the same time was responsible for getting suitable aerial photography. I put pressure on him too. This venture was in its infancy and promises had to be kept if we were to establish ourselves as a viable supplier.

The Army Air Corps supplied our aerial photography but could not now keep pace with our demands. They had other pressures with fisheries patrols, ministerial flights and security duties and we were not a priority. John coaxed and cajoled them, spending days sitting in their hangars trying to convince them to fly, but to little avail. We had tried to reduce the work of the Air Corps by engaging a private firm to fly and photograph thirty towns. The results were poor, with late deliveries and substandard photography. I refused to pay for what was not up to par, reducing the contract price of £30,000 by a third, and I gained a reputation for being difficult to deal with.

'How about leasing our own aircraft and doing the flying ourselves?' John suggested in late 1980, when photography was at a low ebb and the pressure was almost unbearable. 'We could take the camera from the Air Corps and train our own people.'

We both realised that we could no longer rely on outsiders. It was bad enough grappling with the Irish weather without having to deal with the unreliability of suppliers or photography of inferior quality. But retraining scarce staff to manage an aircraft as well as ensuring the willingness of cartographers to fly every time the clouds parted and, most importantly, ensuring that they were available every weekend all year round was a huge undertaking.

John knew that there was no point in approaching me without clear proposals. 'I've asked my people and they'll do it,' he told me. He named three or four staff members and assured me of their enthusiasm. 'I've been making enquiries and found an aircraft and pilots we can lease.' He had estimates of the costs for the one hundred and fifty hours of flying we needed every year. They made sense to me and Muiris agreed.

In 1981 we leased a trundling old McDonald Douglas DC3 that took ages to become airborne and get into position for taking photographs. It was a slow beginning and the aircraft was not really suitable for our purposes, although it had been used for aerial photography in the past. We didn't meet our targets that year.

The following year, we leased a more agile Piper Aztec aircraft from Manchester and the new Ordnance Survey flying programme took off from there, after a shaky start on the first day. John and Andy McGill went to the Great Southern Hotel at Dublin Airport to meet the pilot and inspect the aircraft. Roly, dressed in full captain's uniform, his cap tucked under his arm, presented himself in the hotel lobby. He had the cut and stature of someone about to fly a jumbo jet, not an old plane that could accommodate only three people and an aerial camera. But something else alarmed John. 'Jees, Andy, he's half-blind. We'll all be killed,' he whispered.

Roly was short-sighted and wore extremely thick glasses: this was alarming because Ordnance Survey flying was by visual navigation, sighting on to a distant spot on the landscape and flying directly towards it. To make matters worse, Roly, assuming his captain's role, ordered Andy to carry his bags for him. Andy, lacking due respect for Roly's title, refused and Roly was taken aback. A quick-thinking John decided that the three men needed a cup of strong coffee before the Ordnance Survey's fledgling flying operations came to a shuddering halt. Roly turned out to be an experienced pilot who had flown thousands of hours on all kinds of aircraft and was fully rated for this aircraft. He operated the Aztec for two years before leaving to fly bigger and better aircraft. By that time John and Andy were sorry to see him go.

In 1986 we transferred our operations to Shannon Airport, leased two aircraft and bought two state-of-the-art air-survey cameras. Now there was nobody except ourselves to blame for failure but I knew I could depend on John.

The demand for our services spiralled. County engineers and planners demanded new town maps with updates every few years. Roads engineers wanted faster turnarounds for bypass surveys. We needed more investment in air-survey equipment.

In late 1981 I approached the department for funding for a new air-survey machine. There was a special fund, which had not been fully committed, that was available to support development in information technology. I met two of the senior officials who controlled it late on a dark December day. They were not very sympathetic.

One was a cynic who questioned the need for maps at all. 'All that is needed are the roads,' he muttered dismissively, 'and they were mapped years ago. Anyway haven't you enough equipment up there?'

The second hated making decisions. 'You'll have to send in more documentation and back-up. I can't make a decision on what you have given me.'

I presented my case for two hours. Their negativity did not bother me because I knew that I would secure the money somehow. I adopted Muiris's technique when he met resistance to something he believed in: I repeated the merits of my case over and over, no matter what the objections were. The following January we got the additional machine we needed.

The pressure continued as the demands of a developing Ireland had to be satisfied. I thrived on the excitement of this self-inflicted pressure and spent every hour of the day looking to improve ways of doing things. I constantly examined new developments in computerised mapping, putting pressure on suppliers to introduce further innovation into their systems and implementing the innovations each year. I looked for new ways of increasing revenues and reducing dependence on the exchequer. I never ceased to think about the Ordnance Survey: what had to be done tomorrow, what problems had to be solved, where we would be at the end of the year.

Early on the children often came with me to the office during school holidays. They made friends with some of the other managers and with Harry Crowe, who photocopied their hands. They helped Margaret in the army tuck shop, and ran up and down the corridor of Mountjoy house past the busts of Thomas Colby and Thomas Larcom, making nuisances of themselves. All this gradually stopped as pressures in the office grew. Gradually, too, I forgot to enjoy myself. At first, the travel and the work of developing Ordnance Survey was very exciting. I had to do it, had to get Ireland on the map: every field and ditch, hedge, road and river, every townland and county – perfectly, perfectly into the computer. But I forgot what it was like to stop, to listen, to hear, to see, to feel, to reflect.

As I worked in the garden at weekends and on summer evenings. I no longer stood to ponder on the wonder of the red June sun sinking behind the trees on the canal at Hazelhatch or felt its warmth on my back when it shone across the fields of Colganstown. I no longer noticed how the rising sun cast the shadow of a Scots fir on the wall of the house. The grass had to be cut, the weeds had to be pulled: chore after chore to keep the place in order. Meanwhile, there was always another great challenge in the Ordnance Survey to be considered.

I forgot I was in love or at least I did not have time to show it. I forgot there was such a thing as tenderness. I forgot that it was all right to fall apart and let a loved one see you in your vulnerability and help you out of it and I forgot there was such a thing as working with someone and sharing the burdens. I had no time for that: there was too much to do; the mapping of Ireland had to be finished. It was easier to bring the intensity of work home with me and forget that it affected others too.

I made fewer visits to the country: I couldn't take the time and the trips I undertook were rushed. Anyway, life on the field was easier now. Helicopters had removed much of the drudgery from climbing high and awkward hills: Malachy McVeigh had seen to that. Malachy was an entrepreneur and enthusiast, effervescent with ideas and plans for improving the Ordnance Survey. He had spent his entire career in fieldwork and his commitment to carrying out projects was beyond doubt. His style was individual, very different from the mind that liked to see detailed plans set out on a spreadsheet. He attracted people, getting to know them instantly, their lives and businesses and pastimes, and never ceased to tell all he met of the wonders of the work of the Ordnance Survey. He never forgot a name and people never forgot him. It took me some time to understand his ways but I learned not to query his route to a goal, for he never failed to deliver on a plan.

Malachy was always on the lookout for deals that could promote the work of Ordnance Survey. He seized an opportunity when he met Joe Durnan, a helicopter pilot from Irish Helicopters, in a hotel in the west of Ireland. Joe had been ferrying operators and supplies to the island lighthouses on a monthly basis. Malachy persuaded him and Irish Helicopters to work for Ordnance Survey during their slack times, slinging materials on to mountain-tops and ferrying men between summits. It was cost-effective for Ordnance Survey because we did not have to pay positioning charges.

Malachy spotted an opportunity to organise a charity event for the Cheshire Homes using helicopters. Four Ordnance Survey cartographers were to attempt to climb the four highest peaks in Ireland in one day and set a record for the fastest ascent time. Celtic Helicopters had agreed to ferry the cartographers to the foot of the mountains.

Charles Haughey, who was then Taoiseach, and whose son owned Celtic Helicopters, agreed to launch the event at the Cheshire Home in the Phoenix Park. Malachy was on first-names terms with the Taoiseach, whom he had met when surveying on Inishvickillane off the Kerry coast. He introduced me to him in a rather unorthodox fashion: 'Sir, this is Charlie Haughey. Charlie, this is Captain Richard Kirwan.' Charlie was not impressed and with a regal stare offered me a limp hand. Malachy was untroubled by the gaffe.

In 1989, on one of my rare field trips, I decided to see how the helicopter operations were working. The triangulation work was still in progress and the survey crews were measuring the distances between the islands and hills on the mainland. I went to Inishmurray, an uninhabited island off the Sligo coast. There was nothing there but a few abandoned houses and an ancient ruined monastic settlement. Joe took Malachy and me to the island as heavy squalls swept across and waves lashed the coast. He assured us he would be back unless the weather deteriorated, then left us to battle the gales and make our observations.

I sheltered behind a stone wall as we waited to make contact with the crews on the mainland. The next computer investment was on my mind; bigger, sharper and faster screens and a brand-new air-survey machine with twice the computing power of the previous version. I was about to issue tenders. I hardly noticed when the shadows of the sun racing across the island gave way to lashing rain but when foam from crashing waves blew all about me I was completely distracted from my thoughts of work. Not for a long time had I heard the multitude of sounds of the sea rushing through the rocks or gurgling into little crevices or sweeping from the shore to meet an oncoming wave. I was overcome by sound. I listened as the gale blowing from the Atlantic through stone walls and tufts of grass told its story in its own voice of many sounds. It did not mean to hurt or destroy; it neither loved nor hated; it did not come deliberately. It seemed uncontrolled, yet it wasn't; it merely followed the immutable laws of the cosmos. The rain blew horizontally over the walls, neither loving nor hating, and joined the chorus of sounds in nature: wind, waves and sea.

Joe took us off the island as dusk was setting in and I returned to Dublin to meet further computer suppliers. Malachy drove to Achill Island that night, where he bumped into John Healy, *The Irish Times* political correspondent, who was sitting on a high stool in a bar enjoying a quiet pint. John's distinctive round face and bald head were unmistakable. Malachy shuffled a stool and offered him a drink. 'Not a day for choppers and islands,' said Malachy, addressing both John and the barman.

'Only for surveyors and the insane,' responded the barman, who knew all about Malachy's activities.

John wanted to know more and Malachy obliged, regaling him with stories about the introduction of helicopters into the work of the Ordnance Survey; how the helicopter slings worked and the amount of material they could transport in one load; how dangerous it could be if, at the wrong moment, a sudden gust of wind caught the helicopter on the hilltop; and how the air currents could wreak havoc when the helicopter attempted to land. I have no doubt that Malachy saw the prospect of some publicity in *The Irish Times*. 'I'll tell you. It's no joke when you're leaning out of a chopper strapped in a harness.' He was telling John how he got great photographs of the islands.

A man wearing a cap sideways entered the pub and hurried over to join the conversation. He knew Malachy well and had worked for him, hauling materials up the mountains with his donkeys. He knew every path to the summits, knew to the minute how long it took to climb every hill, could tell at a glance when it was safe to climb and to the minute when clouds would descend and shroud the mountains. He knew how to make the best of a deal as well. 'And you know,' he said to John, 'I got a new suit, drink for a week and the boat fare to England from it.'

'That's it,' said John in a moment of inspiration. 'From Asses to Helicopters. That's the story I will write!' He never managed to write it. John would have seen the irony in such a story. His book *No One Shouted Stop (The Death of an Irish Town)* (1968) was a commentary on emigration from the west of Ireland and the failure of successive governments to set up sustainable infrastructure in the province. Now progress in Ordnance Survey meant less work for the people and forced more of them to seek jobs in far-off fields. It was a pity he did not write it for it was the end of an era, a hundred and fifty years of hauling materials to mountain-tops by hand and by donkey.

15

The Gamble Pays Off

During the development of Ordnance Survey that occurred in the 1970s and early 1980s, the emphasis was on large-scale urban and rural maps. These were essential both for recording changes in the Irish landscape and for preparing for the extension of towns and cities into the green countryside. There was very little attention paid to what we called small-scale maps: those used for tourism, leisure activities and transport. The principal ones included the maps at scales of one inch and a half-inch to the mile. These were topographic maps, showing the coastline, general outlines of cities and towns, road, railway and water networks and the general relief of the landscape. Fields and field patterns, townland and parish boundaries and minute particulars of towns and cities were not shown. The scales were too small for that.

Thomas Colby's teams carried out preparatory work for a series of one-inch maps in the 1830s, when hill sketchers were sent to the countryside to depict the relief, but work on the production of the maps had a chequered history and the first one-inch map did not appear until 1856. One of the most hated figures in nineteenth-century Irish history almost ensured that it never saw the light of day. Charles Trevelyan, the son of an English clergyman, became Assistant Secretary to the British Treasury in 1840. It was his view that the Irish Famine of the 1840s was a 'mechanism for reducing surplus population' and a punishment from God for the perversity and immorality of the Irish. It was he who allowed corn to be exported at the height of the Famine.

Trevelyan was extremely unhappy when he assumed office and discovered that the Irish Ordnance Survey had been set up without written permission from the Treasury. It offended his bureaucratic nature that the paperwork had not been completed. This was a strange attitude to take, as a select committee of the House of Commons had agreed to its establishment and it had been in existence for fifteen years. Trevelyan promptly made it clear that no new map series should be undertaken, nor new methods of surveying introduced, without his express permission. He was also vehemently opposed to public moneys being spent when, in his opinion, a one-inch map series could be produced by private enterprise if there was a popular demand for it. He overlooked the important fact that it was the military and police who most needed the one-inch maps to keep the perverse and immoral Irish in check.

Colby was not a man to let setbacks like this deflect him from his purpose. He had overcome many such setbacks since 1823 and seized an unexpected opportunity to continue with his preparations when, in 1844, one of Trevelyan's staff issued a letter with an un-intended meaning. It stated that the Ordnance Survey should continue with the work of hill sketching, not realising that its only purpose was for publishing the one-inch maps. The final sanction to publish the maps did not come until 1851 and all the one-inch maps were published by the close of 1862.

Ireland would have been the poorer without such a map. Six-inch maps covered too small an area to convey a general view of the landscape. The one-inch showed the inter-connectedness between places, towns and villages. Hill shading, hachuring and colour brought the maps to life.

This was the map that first spoke to me, that invited me as a boy to wander the roads of County Waterford and dream and associate memories and incidents with particular places. And whenever, later in life, I looked at the map at the townland of Ballinattin near Tramore, just off the main Waterford road, one of my father's stories came vividly to my mind. It was a place to which we would regularly cycle in summer, along the narrow road through Ballinattin that overlooked the back strand and sandhills of Tramore. When the tide was out we would trace with our eyes the myriad meandering streams that flowed through the sand to the sea. But first, we stopped at the red-brick memorial to the IRA volunteers killed during the War of Independence. My father would, without fail, take off his cap and pray for the fallen. 'Great men,' he would say. Every time we passed the memorial I would ask him about Nicky Whittle and the Pickardstown ambush that took place in 1921.

'Look,' he would say pointing down the hill to the junction of the road we were on and the main Waterford to Tramore road. 'That's the spot.' He talked as if he had been there. 'Two brave men died and two were wounded. They didn't have a chance.'

Fifty volunteers from the IRA, many very poorly armed, had hidden in the ditches to ambush a British Army convoy of four lorries travelling from Waterford to reinforce the Tramore RIC barracks, which had come under attack. The volunteers were outnumbered and the ambush went badly wrong. 'Look,' my father would say, pointing to the fields sloping upwards from where we stood. 'The poor devils had to climb through the bare fields to escape. There was no cover for them.'

Nicky Whittle, my father's first cousin, was shot three times and escaped only by crawling through the briar ditches. The IRA declared him 'officially' dead, to keep the British from searching for him. His family went into deep mourning, held a funeral and buried his coffin. My heart was always in my mouth at this stage, no matter how often my father related the story. I had become part of the drama, worrying that the British might snatch the coffin and open it to find nothing inside. 'What would they have done?' I always asked. Nicky emerged into the limelight again with Irish independence and became the Sinn Féin director of elections for Waterford in 1923.

The one-inch maps became a record of an era because, apart from the maps of Dublin, Cork and Killarney, they were never significantly updated after Irish independence. Nothing more than the original names for the railways were changed or deleted. Names like 'The Great Southern and Western Railway' and 'Waterford and Maryborough Branch' were no longer used since Iarnród Éireann had assumed responsibility for the railways. Perhaps that is how the one-inch maps retained their magic. The hands of engravers, like James Duncan, the chief engraver at Ordnance Survey, who between 1827 and 1866 pioneered many of the new ways of depicting the landscape on copper plates, are forever present in the lines, as are the records of the field surveyors, offering clues to the past and enticing the reader to journey into the heart of the map.

Inishmurray, the island on which the voices of the wind and rain opened another world to me, had a lot to tell. It was a holy island. The one-inch map showed a monastic settlement and holy places with the names printed beside them: St Molaise's chapel, St Molaise's monastery, and the old well, Tobernaroragh, all from Ireland's golden age of Celtic monasticism. The curious reader of the map might look beyond it at the six-inch map for more information. This showed the locations of other churches and their various names; Temple Molaise or Templenabar, Temple Murray or Templeterraman. There were the flat rocks or penitential slabs of the saints, Lachta Patrick, Lachta Colmkille. There were altar stones and sites of crosses such as Crossmore, presumably the largest cross of the Island.

And the six-inch map might make one curious about landmarks such as Saint Colmcille's stone. Legend tells that Saint Molaise, who founded the monastery in the sixth century, was Colmcille's confessor and that Colmcille fled there to seek forgiveness after the Battle of the Books at Cúl Dreimhne in County Sligo. Saint Colmcille had copied a psalter while he was a guest of Saint Finian and had refused to give it up to the owner. The high king decreed that he should give the copy to the owner – 'to every cow its calf and to every book its copy' – a judgement that may have been the first ever copyright ruling. But Colmcille refused to give back the copy. The high king and Colmcille fought a battle on the slopes of Ben Bulben, where it is said, thousands were killed. Colmcille won the battle and the high king was forced to reverse his ruling. Colmcille, overcome with remorse at the slaughter of so many people, confessed his sins to Molaise and went into exile to the island of Iona. He vowed never to return to Ireland.

The one-inch map also provided clues to Inishmurray's later

history. The nineteenth-century schoolhouse was marked, as well as the ruined houses I had seen on my visit. Clashymore harbour, from where monks and later islanders crossed to the mainland, was shown on the sheltered side of the island. It was probably from here that the last of the islanders left in 1948.

The half-inch to a mile series was born against a background of the First World War, the possibility of Home Rule for Ireland, the 1916 Rising and the fading glory of Ordnance Survey. It was first published between 1912 and 1918. The series was derived from the original one-inch maps and was intended to sit side-by-side with them. But the coming of Irish independence changed all that. As was the case with the large-scale maps, the state did not see the need to update the series and in any event it could not afford to. They were left to survive as they were until the one-inch series was finally abandoned and the half-inch updated in the 1940s.

The half-inch series became the public face and workhorse of the Ordnance Survey. Its plain brownish cover with Ireland's outline and the county's name was unmistakable. It was a cover that would not change for forty years, dull and uninviting, in keeping with the austerity of the times. Generations, especially geography students, knew it as 'the Ordnance Survey map', as it was a mandatory part of the geography syllabus in secondary schools. There was a compulsory map-reading question on the state examination papers, an easy one for those who liked maps.

The half-inch map did not speak to me in the same way as the one-inch. It was vivid and sharp, clearly showing roads and rivers and railways and illustrating the relief of the landscape by means of coloured layers. It got me to the hills I had to climb and down the narrow roads to their lower slopes. But it never had the depth of information of the one-inch. Inishmurray was just an island off the Sligo coast. There was no hint of its history, or that it was ever inhabited. Maybe I was biased because my childhood dreams were buried in the one-inch map.

In 1965 Gerry Madden and his committee recognised that a new small-scale map series was urgently needed. The committee proposed a new metric scale of 1:50,000, one metre on the map representing 50,000 metres on the ground. We called it the 50,000 or 'fifty-thou'. An all-Ireland 50,000 series was subsequently agreed with our counterparts in Northern Ireland. The Ordnance Survey of Northern Ireland had been adequately funded since the end of the Second World War to update both their large-and small-scale maps. The worsening security situation of the 1970s gave them a greater impetus to get things done. We had never taken the project forward because of lack of resources and, in any event, our priority was the large-scale maps.

By the early 1980s complaints about the inadequate state of our small-scale maps became more frequent and adverse comparisons were being made between our maps and the Northern maps. We would have to make a start somehow. We needed strong, visible support, not just anonymous verbal complaints that were never followed up in writing. Muiris Walsh used to say that the complaints file on the state of topographic mapping was the thinnest in the office and that there was little point in approaching the department or indeed the minister without the evidence he needed to obtain their sanction for spending in this area.

In 1989, almost twenty-five years after the original recom-mendation, Muiris convened a committee of interested parties in a push to gather support for the project. Hill-walkers, Bord Fáilte, the emergency and security services and geography teachers were represented. We had produced and published a preliminary map of the Magillicuddy Reeks at the new scale of 50,000 in 1998. There was widespread praise for it and strong support for our mapping the rest of the country, something we already expected. We ambitiously proposed that a new series should be completed in five years. The committee's recommendations were accepted by the Department of Finance and we published a second map, of the Slieve Bloom Mountains in the midlands, in 1989.

The maps were different in style from the Northern Irish ones. Time had moved on and we had the benefit of computer cartography. Most of the Northern Irish 50,000 maps were manually drawn to a jointly-agreed specification which was really only a variation on the look and style of the old half-inch. I did not want to go down that route. I wanted to make use of the power of computer cartography. We came up with a new design, a draft style, because computer cartography was still in its infancy in 1989. We wanted to bring out a preliminary series so that interested parties could provide suggestions and input to the final content. Most users, with the notable exception of teachers, were pleased with the preliminary maps. The teachers thought it was impossible to read the contours and teach the geography syllabus with them. But it was a start.

The way these maps were produced was a complete break with the one-inch and half-inch maps of the nineteenth century. It was the first time that a completely new map of the country had been produced using aerial photography, computers and a small amount of fieldwork.

At the beginning it was neither easy nor fast. Only five maps were published by the end of 1991: Magillicuddy Reeks, Wicklow Mountains, Slieve Blooms, Ben Bulben and Collooney in County Sligo. Our five-year plan was a distant dream: at this rate it could take forty years to publish the entire country. There were teething problems with the drawing software and the cartographers had to be trained; nor did it help that the managers were from the traditional school of manual cartography.

I had been involved only on the fringes of this work, principally with the aerial photography and the contour production, being too busy to be enmeshed in the detail. But by 1991 I was becoming concerned with the quality and the slow progress. Sheet 16 of County Sligo had just been published and I suppose I took a greater interest in it because Inishmurray appeared on it.

I was alarmed when I took a close look. The map was not a pretty sight, even for a preliminary version. A contour was misplaced and drawn in the sea. The names of towns were inconsistent in size, with some bigger towns having smaller text than small villages. Worst of all, a major causeway connecting Dorrins Strand to Coney Island was missing. I arranged to have the map quickly withdrawn from circulation.

I put my concerns to Muiris when we met one day in his wood-panelled office where the silver coffee pot gurgled on. I volunteered to take a more direct interest in the 50,000. But I insisted that there had to be managerial changes if I was to assume direct responsibility. 'Something has to be done,' I said. 'This project has too high a profile.'

He was concerned because I had enough on my plate but reluctantly let me have my way. 'What changes do you want?' he asked.

He was taken aback when I said I wanted Harry McDermott and Malachy McVeigh to manage the programme. He had every right to be. Harry and Malachy had spent their working lives on fieldwork and had no experience of cartographic production.

'Malachy has the energy and enthusiasm,' I told him, 'and Harry is level-headed. He will give Malachy the benefit of his management experience.'

I got my way and they took over in 1992. Putting inexperienced men in charge was a gamble. Five new maps were published that year. It rose to ten the following year, then to twelve.

The first year with the new team was not without incident. Malachy had organised a launch of the map of the Dingle peninsula in Dingle. Muiris and Malachy travelled early to ensure that everything was in order. I was to follow on after a meeting in Ennis with the county engineer. My heart missed a beat in panic as I crossed the Shannon Estuary to Tarbert on the ferry. Had we printed placenames in Irish on the map? I couldn't remember. After all this was a map of a Gaeltacht district.

Mobile phones were in their infancy. They were big, heavy and cumbersome and the signal coverage was problematic, especially in rural parts. Luckily I managed to reach Malachy. 'Are the placenames on the map in Irish?' I asked him.

There was a pause followed by a deadly silence and the lapping of the Shannon on the boat turned to a roar.

'I don't know,' replied a subdued voice eventually.

'Well, would you go and see?' I said, although not quite in those polite words.

There was another pause and eventually an even more subdued voice said, 'No, they are not.'

At that point I lost the reception on my phone and had to drive on to Dingle wondering how we could solve this problem.

There was a potential disaster in the making – a Fianna Fáil senator launching an Ordnance Survey map of a Gaeltacht area in Dingle with no Irish on it. We also expected reporters from local and national newspapers to attend.

I need not have worried. Malachy knew Senator Maurice Fitzgerald and had a quiet word with him to explain that this was just a preliminary map. Ordnance Survey wanted to get the opinions of the locals on how it could be improved and especially wanted their input on the proper Irish versions of placenames. The senator understood and during his address noted that there were no placenames in Irish on the map. He paused for a dreadful ten seconds that seemed a lifetime to me before explaining that the Ordnance Survey had not yet printed Irish placenames because they wanted the help of the people of Dingle and west Kerry. The launch was reported in *The Irish Times* the following day and there was no mention of the absence of Irish. The definitive version of the map had an abundance of Irish placenames.

It was coincidental that new automated cartography systems were coming on stream about the time Harry and Malachy took over the 50,000. We redesigned the series and published a full-colour edition. The first map with the new design was published in 1993. We named the series the 'Discovery Series' and the maps were widely acclaimed by long-suffering map users.

The Discovery Series was completely different in style and looks from the Northern Ireland 50,000 maps, even though they were to be part of the same series. To my eye – and no doubt I was biased – ours was more attractive, softer in colour and easier to read. We took the first map to a meeting with our Northern counterparts. Muiris, Malachy and I were palpably proud of the new product when Muiris introduced it to his opposite number but he did not like it – not the colours nor the layout nor the design. There was always a certain rivalry between the two organisations despite an excellent working relationship. Muiris might have expected some criticism because the Discovery Series had not been produced to the agreed specification but he was shocked at the strength of the criticism, as were Malachy and I. There was no further comment that day but within a few years the Ordnance Survey of Northern Ireland had redesigned their maps to be closer in looks to ours!

The Discovery Series became our flagship product. The gamble with Malachy and Harry had paid dividends.

16

Too Busy to Laugh

The insights I received from the wind and sea at Inishmurray were the last I had for a long time. I became far too busy for introspection as the 1990s began, and willingly allowed my mind to become numbed by the demands of productivity and the gruelling work of getting things done. My dream that Ireland's map-makers should once again be acclaimed as innovators and that the country's mapping be restored to its former glory became an obsession and I, like Muiris Walsh, became the Ordnance Survey. Every aspect of the mapping programme had to be kept going, despite the lack of recruitment and the loss of cartographers. I had to be part of it all: urban, tourist and rural maps, contract mapping, changing work practices and new technology.

The large-scale rural mapping programme worried me most of all. It had to be slow because twenty thousand maps of the country had to be revised in the field. We were looking at a twenty-year programme. I was pushing hard to see how we could reduce this to a reasonable figure when John Danaher surfaced with a really radical proposal: to abandon Colby's maps altogether and remap the country using aerial photography, despite having already updated about four counties. Initially, I dismissed the idea as crazy. At least with Colby's maps we only had to add the new and delete the old. No other country in the world had tried such an approach, not even Britain, where there were greater resources and where similar mapping existed.

But John persisted. He knew that if he was to 'keep coming

down the hall to that man' with any hope of success he would have to have every fact to hand and there was no room for fuzzy thinking. I considered his idea, revised his estimates of cost, time and people upwards and agreed that what could be achieved looked good. From now on, rural mapping, like the urban and tourist series, would be carried out using aerial photography. Now we had the right technology. In 1992, technology developed for the American military was declassified and made available to the civilian mapping world. Polaroid sunglasses and computer screens replaced cumbersome air-surveys machines and operators could now view digital aerial photography in three dimensions. Suddenly everything became faster and less expensive. It took four years fully to effect the change in mapping methods and during this time the Irish Ordnance Survey became the world's biggest user of this equipment, apart from the military.

The unions resisted the changes, principally because one hundred people had to be reassigned from field to office duties. I was glad that Muiris was there to negotiate with the unions as it was not a situation in which I was ever happy. He also negotiated with the politicians, every one of whom wanted to set up a regional office in his or her own constituency to accommodate displaced surveyors.

I often wondered about the men who had looked sceptically at the original air-survey machines in 1965. What would they say now if they saw what was happening? Would they decry the decadence of the age, youngsters with fancy sunglasses looking at aerial photography on screens? Would the men of Colby's era be horrified to see these machines replacing their craftsmen skills of penmanship and copperplate engraving, producing a map in a week rather than months? If so, they would not be alone. The conservative world of the aerial-survey industry looked on sceptically. Until now theirs had been a world of incremental changes. But within five years of our having purchased our first screen and pair of sunglasses the world had moved on and nobody would countenance using the older equipment again.

Colby's era was gone and the final chapter of the modern remapping of Ireland had opened. By 2004 every map sold had been produced from aerial photography. That Colby's legacy was gone did not cost me a thought, if we could remap Ireland faster without it. The old maps became merely references to the past, confined to libraries and archives. Older people still talked with nostalgia about the six-inch map and the men who made it, acknowledging their pioneering spirit and artistry. The new map-makers were too young to know. Theirs was a virtual landscape they might never set foot on.

The pressures of work grew, as air-survey technology continued to develop. Every few months there was a new development. New digital air-survey cameras came on the market which could help further in achieving my goal of a remapped Ireland. I had to explore them all.

Travelling took on new proportions: trips throughout Ireland, Europe and America; a day at a meeting here and another there; a lecture here and another there; a launch and reception somewhere else; a hotel for a night; an aeroplane or hired car to the next destination. Sometimes when I woke at night I did not know which city I was in. In 1997 I visited seven American cities in eight days, with meetings in Boston, Atlanta, Denver, San Diego, Denver again, Sacramento and Redlands in California. I woke up in Sacramento, not knowing where I was or which side of the room the bathroom was on and it took me five minutes to get my bearings.

I was not alone in my attitude to life. Ireland was surging forward in a frenzy of unbridled development and most of its people were riding the crest of the wave. Consumerism was rife with virtually no unemployment, rising salaries and people going into serious debt to buy homes and second homes. Few gave any thought to the ultimate cost.

In 1996 Muiris retired and I became the Director of Operations, a new title for Muiris's job. I took on his burden and kept most of my own. There were further meetings with sceptical bureaucrats. There was endless report-writing and many presentations to make and meetings to organise to gain funding and approval to carry out obvious and necessary projects. At times it seemed it would be easier – and far more rewarding – to climb every mountain in Ireland twice in winter and walk every road barefoot than to deal with these people. But I was not to be beaten or subdued. I studied part-time for a master's degree in management science at the Irish Management Institute and I began to put my stamp on the Ordnance Survey, initiating a process of transforming it into a commercial state body.

Ordnance Survey skimmed along the waves of progress and by 1997 was recognised as a leader in world mapping technologies. The challenges of my office and technology kept my adrenalin pumping. I kept raising the bar for every mapping programme. The urban map programme had been completed but I wanted all the maps to be revised at least once a year – a difficult target to achieve because there was so much change. The manager in charge thought a five-year programme was achievable and on being pushed was prepared to concede a three-year programme. He complained to one of his peers, 'That man down the corridor is totally unreasonable.' I was having none of it. It took him two years to achieve the one-year revision cycle I had set my sights on.

In my push to get things done I forgot about my own life, which undeniably became seriously imbalanced. I brought frustrations and creeping tiredness from work and travel home at weekends. I was hard to live with and intolerant of anything that upset my routine. The work plagued my mind, even in the sunshine of my garden. The next problem and the next challenge were always with me.

Since childhood I had never liked talking about or tackling personal issues. I shied away from people and girlfriends when they became intimate and wanted to know more about me. I would not let them look into my eyes and see my discomfort and vulnerability. Neither would I look into theirs, lest I might have to answer their searching questions. Now I had the perfect reason for not addressing deeper issues: I was too busy. I shunned human warmth and kept personal dealings at superficial levels.

From the middle of the 1980s I took few holidays and in 1997 I took none. I let Eilís and the children fly to Jersey while I went to work and studied for my master's degree in my spare time. I drew the townland boundaries on the maps and I thought they were for ever: the centre of the fences, centre of the roads, centre of streams, invisible boundaries through open fields. That was my life: getting boundaries on the map and making them permanent. I failed to acknowledge that my own boundaries had become more entrenched.

Perhaps I should have paid more attention to the Rye river, the centre of which formed the townland boundary between Nicholstown and Ferns near our first home. It would have shown me that boundaries could be penetrated by the softly flowing water caressing the river bank, nibbling away ever so gently at the hard clay. I could have seen that, regardless of boundaries, a river could alter its narrow course over time, meander softly and slowly through the countryside and still reach the sea, its final destination.

Or I might have paid more attention to the sea flowing past the seaweed-covered rocks on to the sands at Kilfarrasy, a beach on my childhood map. I should have seen it shape-shifting every day, altering the line of debris that marked the high-water mark, another boundary I had chased to draw on the maps of the Ordnance Survey.

Or I could have listened to the message in the storm on Inishmurray. Was it advising me? Was it telling me that the force of a storm was sometimes needed to knock stubborn boulders from a boundary blocking the way, in order to expose hidden terrain?

Whatever the messages of the river or the sea or the storm, I did not comprehend them. I was in my own invincible ambitious world and would not be deterred. My colleagues told me I looked tired and should take some time off. Some dared to suggest that I delegate more. I dismissed their advice without a thought. I was all right. I just needed to push a little more and the tourist maps would be finished and recognised as a masterpiece of computer cartography. The rural map programme was progressing; it only needed a little more of my time. Every morning I put on the suit and tie and took the briefcase to the office. Another day behind the office door, another day behind the mahogany desk, another day too busy to laugh.

But thick, grey, turbulent storm clouds were gathering, floating and hovering menacingly, threateningly, above my head. At times they almost touched it. They had warned me and I had ignored them. In November 1998, their patience exhausted, the storm clouds burst and the ocean came crashing into my unbalanced world. The galloping white horses of the sea ripped apart my well-constructed boundaries.

17

The Font of Enlightenment

Séamus, a friend, noticed what I could not see and after a meeting one day invited me for coffee. He never mentioned my work rate but suggested that I attend another course at the Irish Management Institute called 'Untap Your True Potential'. This sounded good to me. Séamus said it was about exploring ways of improving management skills but the methods used were a little unorthodox. Much of the course involved hypnosis. I immediately told him to forget it because I did not believe in such rubbish. Then he said something that made me laugh. 'Look, at least it will do you no harm. If nothing else you'll be lying on your back for a few days and can have a kip at the state's expense.' That softened me and I registered for the course, although with a very sceptical mind.

Larry McMahon, a likeable man who drew people easily into his confidence, was conducting the course. The morning sessions were given over to interpersonal skills and the hypnosis sessions took place in the afternoons. Larry reassured us that it was not about seeing leprechauns and that he had no power over people in a trance. The subconscious mind would allow only what was in a person's interest to happen. He sounded convincing but I and others in the class remained sceptical. One was not convinced at all – or perhaps he was fearful of what he might uncover – and decided to resist Larry's attempts to put him into a trance.

Larry said he would address our subconscious minds and ask them to bring about changes appropriate to us, such as stress management, improvement in creativity and time management.

He encouraged us to form our internal intentions relating to our needs which he would address in the trance. My intention was improvement in management.

He started: a soft, pleasant voice asking us to breathe deeply, then hold our breath for a while. He continued; asking us to notice what we could see in our mind's eye. On and on he went in the same gentle voice, asking us to be aware of our hands, feet, legs and bodies. I lay there impassively and remember saying to myself, 'Yeah, yeah, Larry, where is this rubbish going to lead?' But then I became aware of a strong clicking sensation in my mind. It sounded and felt as if someone was flicking hundreds of stiff pages of a book backwards. I had no control over it; I was just aware, a witness to an event. Larry's serene voice drifted in and out of my consciousness and the flicking continued.

I slowly came too after the trance and dragged myself from the floor. I was feeling groggy and sad and my heart was racing. I thought we had been lying there for ten minutes at most but the clock said that an hour had passed.

Larry was curious about our experiences but there was no compulsion on us to elaborate on what happened during the trance. When I told him about my heart racing, he said I either had a heart problem or something big was happening to me. I knew I did not have a heart problem. And as I was telling him about my feelings I remembered a white terrier pup I had when I was young. My father had him put down because he had bitten me but he told me that he had given him away. I was only five or six years old and I spent weeks after school sitting by an empty dog box, imagining the pup was still lying there licking my hand. Afterwards, I thought it was strange that nothing about the Ordnance Survey or my work crossed my mind during my time in the trance.

The class went for coffee and I went for a walk. I suddenly broke down and hegged and hegged with sadness for the pup. And then the sudden death of my father when I was thirteen hit me. I had never grieved either for the pup or for my father. I had never spoken about them or my great friend, my grandfather, after their deaths. They had abandoned me when I was young and there had been no one to take their place. I never confided my loneliness to another person, not even to my mother. My father went to visit his sister one night and never came home. My childless aunt came to my bedroom late that night and told me that I was now the man of the house and had to be brave and should not cry. I did not grieve, I did not cry. At Butlerstown cemetery I listlessly watched his coffin being lowered into the ground as the priest blessed it with holy water and recited the prayers of committal. My mother and relatives stood beside the hole in the earth and the mound of clay dug from it, cried silently and wiped away the tears with white handkerchiefs. Clay, shovelled on the coffin by the gravedigger made a hollow, thudding sound that reverberated in the cold February air. I stood there, motionless and emotionless, and recited a lonely, hollow prayer. I knew that now everything depended on my bravery.

Now, thirty-six years later, alone in the grounds of the Irish Management Institute, the grief came pouring out. Now I could talk about it all. The result of this trance was not part of my plan for my life and certainly not part of the plan I had for this course.

That night, as I drove home in a daze, overwhelmed by all that had happened in one short hour I felt that an unknown burden had lifted from me. I remember going into the house and throwing my arms around Eilís, saying, 'I'm back. I'm back.' It was a spontaneous action. I did not know why I was saying it. All I knew was that my carefully ordered world was in chaos.

The next day was Saturday and I was glad not to have to go to work. I felt like a wet rag and had an aching throb in the back of my neck. I though it might have been some old injury healing as a result of Larry's work. The pain was draining me and I had to lie down, something I would never dream of doing on a Saturday. It was always a busy day: the children had to be delivered somewhere, there was a match to go to or the garden had to be taken care of. That Saturday Eilís had to take the boys to their match.

About three in the afternoon I got up to go to the shops in Celbridge. But I found myself involuntarily driving a further four miles to Maynooth and travelling on into Saint Patrick's College, a seminary for the priesthood. I had been there before, had visited the garden and walked the grounds of the college, but not for two years now, and the garden had not crossed my mind since then. And now I had no earthly reason to go there, no logical grounds to drive in the rain to visit a water garden, let alone get soaking wet walking from the car to the garden.

I had never thought much about this garden except that it was pleasant. It had been built a few years previously to mark the bicentenary of the founding of the seminary in 1795. Pools of water and irregular rocks and slabs of limestone sought to bring the wildness and beauty of the Burren landscape here. But for me, the wild spirit of the garden had been suppressed by the austerity of the tall, grey seminary buildings surrounding it on three sides.

On this Saturday afternoon I left the car and walked slowly to the garden. Soft drizzle, almost suspended in the air, fell delicately and lightly and touched earth, rock and water without a sound. The narrow neo-Gothic windows of the four-storey buildings were firmly closed against the wetness of the day and neither saint nor sinner peered through the small glass panes. The place was still and deserted.

I walked, oblivious to the rain, past young trees and shrubs, each one labelled with its Latin name and the region of the world to which it was native – Morus alba, white mulberry, China; Pinus pinea, stone pine, Southern Europe and Asia Minor; Plantanus orientalis, oriental plane, east of Kashmir and south-east Europe. I came to a place with three fountains set at the edge of the main body of water. They were named the fonts of Procreation, Pain and Power, each simply a large rock from which water spouted before it gently dribbled over the sides. I was drawn to each in turn and stood unable to move, held by an energy I could not understand. I could do nothing but contemplate the mysteries of the three fonts; how pain and power followed on from the joyful act of procreation; how power and pain were intertwined and how they impinged on my life and every other life. When the mysterious energy dissipated, I blessed myself with the water of the fonts and was released.

I walked on over a narrow waterway to three other fonts, Faith, Truth and Wisdom, and was halted again by the mysterious energy.

I contemplated the sureness that is in Faith, the liberation that is in Truth and the ease that is in Wisdom. I blessed myself again with the water from each of the fonts and the energy released me once more. I walked on to the final fountain, the font of Enlightenment, a tall, smooth, slender limestone rock that reached to the heavens. Unlike the others, a strong flow of water spouted from it and a distance of water separated it from the edge of the pool. I meditated on the great mystery of Enlightenment but I was unable to reach its waters.

Why was I here? Why was I going through such a ritual when I had long ago replaced fervour for religion by my busy lifestyle and important work? What was this ritual all about? How was it going to affect my life, tomorrow, next year, years hence? When would I get the answers? I asked myself the questions over and over again but no answers came.

I learned later that this was a Biblical garden, something which should have been obvious to me from the beginning. Everything in the garden was a representation of some stage of humanity's development, from the nadir of existence to eventual salvation. The entrance to the garden was through a barren wasteland, a dried-up river bed strewn with tree trunks and rotting trees. Nearby, clumps of thistles represented the trials of human life. The fonts represented the stages of development through which humankind must pass before reaching the goal of salvation.

There was another route through the garden, a path by which humanity might have avoided the trials and tribulations of life and approached Enlightenment directly. But that path was inaccessible. A bridge on it was damaged, shattered by the Original Sin of Adam and Eve. I also discovered that the font of Enlightenment was accessible by walking on flagstones slightly beneath the water's surface. But on that visit, I was not meant to notice them.

I returned home puzzled but not surprised by what had happened. Eilís wondered how I had got so wet. She said nothing when I told her about my adventures. I think she thought I had taken leave of my senses. I was in a daze for the next week. My whole outlook on life was changing. The Ordnance Survey seemed merely a part of my life now. I began to feel that perhaps I was gifted with the power to hypnotise and could be a channel for healing. I discussed it at home and tentatively experimented a little. It worked. Then Mary, our second eldest daughter, developed a serious throat problem shortly before her final university examinations. The doctor thought she might not be able to sit the examinations. Eilís and Mary agreed that I should try healing her through hypnosis. Within a day her throat had healed.

Over the following months I began once more to take time off work. A new creative spirit in me wanted to go into the garden and work with the soil. I no longer brought home the problems and crises of the office. I could relax and notice the grass and shrubs and everything that grew without seeing some office torment tangled in them. For the first time in years I had the urge to build water features. I built three major features, digging and constructing each by hand. The first was a double pool, where water from one pool tumbled loudly into the next. Yet there was silence in its turbulence. The next was quieter, with dribbling waterfalls where the water continually chattered before finding its way to another pool. The final pond, silent and still, reflected tall artichokes nearby in the dwindling light of evening time and brought tranquillity to the garden.

Tiny flat stones in the stream that separated our townland, Commons, from Colganstown wanted to be part of my creation. Black from years of being submerged in flowing water, they invited me to lift them from their resting place and make them part of the low walls in the pools. I picked thousands and laid them one on the other so they looked as if they had always been there. A pathway through a rockery adjoining the still pond directly faced the place on the horizon where the sun set at winter solstice: a coincidence. That year I watched the sun slowly draw the pagan year and the second millennium to a close. It descended slowly through the cloudless sky and sank behind the distant trees in the townlands beyond Colganstown. Its last ray dipped under the branch of a hawthorn tree at the edge of our stream and rested on the pathway. I bid farewell to the closing year and gave thanks for all the growth the sun had brought.

Larry McMahon's workshop had a lasting effect and I began to feel more relaxed in relation to work. Nobody knew about my experiences. However, I was not prepared to water down my vision for the modernisation of Ordnance Survey or the completion of the remapping programme. I pressed on with the same impatience as before. Meetings were as lively as ever and people were still reluctant to 'come down the corridor' without considered proposals or reports. I was still looking for new ways to do things, especially some means of speeding up the rural mapping programme.

I began to explore the possibility of outsourcing the work to private companies. This would be a new departure for Ordnance Survey and would need the agreement of unions. It was a delicate issue which the unions would naturally oppose. It could not be done overnight. We now employed a full-time human resource manager to shoulder some of the burden. We bounced the idea off the senior people like John Danaher and broached it with the unions. The idea would have to mature in everyone's mind before serious discussions could take place. I also had to be certain that private companies could meet the standards required and that their costs were reasonable. It took three years of discussions before the unions agreed to outsource most of the remaining rural mapping work. When it was completed, Ordnance Survey cartographers would do continuous updates. In 2002 we awarded contracts for part of the remapping of rural Ireland to three companies, one in Northern Ireland, one in Scotland and one in India. We supplied the aerial photography and they drew the maps. The biggest contract was with the Indian company and in time it became our only supplier. It was ironic that Ireland had assisted India at the beginning of its great triangulation and mapping programmes by giving it Colby's measuring bars. Now India was using its skills to help Ireland to complete its mapping programme.

A year to the month after I attended Larry McMahon's workshop I attended another which accidentally came my way. It was a weekend workshop called 'What Colour Is Your Parachute?' It was both a weekend of reflection and a sort of practical guide to high-lighting one's strengths. In many ways it was designed for people seeking a change in career or lifestyle and many of the methods used to highlight strengths were empirical. Towards the end we were given an exercise which consisted of answering various questions and ticking boxes. This was to give us a fix on the order of our strengths. I expected management to top my list; after all I had spent my life in management and had achieved a lot. But it was only my second strength – top of the list was healing. I was astounded.

I expressed my amazement to the tutor and she said I should try hands-on healing, particularly Reiki. I didn't know what she was talking about. The following Wednesday I was in Dublin and, between meetings, wandered into a bookshop in Dawson Street. Staring out at me was a book on Reiki written by Teresa Collins, a woman from Cork. I managed to get her phone number and rang her immediately. 'That's strange,' Teresa said. 'I have a course starting this weekend and there's one vacancy.' I discussed it with Eilís and she encouraged me to attend. During the following year I travelled to Cork and undertook further courses in the various levels of Reiki until I became a Reiki master.

It was a strange transformation for someone who had adopted a thoroughly logical approach to life. With me, everything had to be reasoned, or so I always believed. Now here I was engaging in a form of hands-on-healing and working from an intuitive place in myself.

18

EVERYTHING HAS ROOTS

In 2005 I decided to return to my roots and create some balance between my past and all that I had become. I would spend more time in Waterford, the place where I had made my first map, that imaginary map of the fabric of the city in which I was born. I had been back over the years on fleeting trips to see my mother but had no time to savour the changing city or dwell on its past.

I could once again see the city's uniqueness and realise why it was worth making my original mental map. As I approached Sallypark, under the shadow of Mount Misery, I noted the city of the single spire, that of the Protestant Cathedral. The belfries of all the other churches, those in use and those in ruin, peeped above the grey rooflines of the city. The buildings of Ballybricken, still as they were in the 1960s, stood proudly above the city centre. The cattle fairs once held there were history, as were the rallies of Daniel O'Connell, Charles Stewart Parnell and John Redmond of the Irish Parliamentary Party. But the river flowed as it always did, rising with the incoming tide, exposing its mud banks at the quay walls as it ebbed.

The city had changed, not in its street layout, but in the way it conducted its business. The Quay had become a place of leisure, a place to stroll along the banks of the Suir or to park a car. The wartime concrete ship near the bridge which had fascinated me so many years before had been removed lest it offend the eye. I had not noticed that it was gone until now. The moorings, or hulks as the were called, for the great ships of the past had been replaced by floating pontoons for yachts and leisure craft. Only those who owned the boats now had access to the moorings.

I had to walk slowly through the city and anchor myself once more. My childhood map was returning. Images, memories, emotions were not dead as I had thought, just dormant. Stephen Street was where I attended primary school. The school was still there but the brothers in black cassocks were no more; they were too few and too old to go on teaching. It was here that my map of Waterford first expanded to encompass the whole of Ireland, an Ireland divided, as the brothers had told us. That shiny map, yellowing and cracked from age, hung on the classroom wall – an outline map of the whole country showing the principal towns, mountains and rivers. Aided by prompts or roars or threats from the brothers, I learned from it the names of the counties and the rivers that flowed through them. I remember vowing some day that I would travel Ireland, explore each county and get to know every acre of the land.

It was in the classrooms of Stephen Street School that I got my true learning from those great brothers like Virgil and Bernard, who illuminated my mind and set it on fire with Mathematics, Geometry, History and Opera. There too was Edmund the artist, who mesmerised me as he made a fretsaw waltz through wood and bring the inanimate to life. Those brothers had fuelled my enquiring mind, made me want to look under stones and rocks to see what was there. Theirs was a spirit of adventure. They distilled simplicity from the complex and made the impossible look achievable.

I also encountered the darker side, its pain and suffering. I lived on the edge of fear. John loved his cane – the swish, the crack, the red hands and fingers, the agony on our faces. Any excuse would do. Daniel, the left-handed demon, was quick to connect his hand with the face or the back of the neck, impatient for a palm to feel the force of his leather strap laced with coins. The cutting edge of Maurice's tongue was even more incisive. Fear ensured that you remembered the counties through which the three sisters, the Nore, the Barrow and the Suir, cut before they merged and flowed into the sea. And there was the brother who showed us how to be cynical about life by casting derision on the mayor who, during his annual visit, spoke of the untidiness of the city and the 'literature' strewn around it. Yes, we learned and remembered from that. We even had to write an essay about it.

There were those who told me I was not good enough, who poured scorn on me when they did not like my English compositions or when my spellings were incorrect or when they just wanted to demonstrate their authority. And sadly I believed them.

But all this was a harmless memory now; the number of brothers was dwindling. I saw their vulnerability and frailty when Brother Joseph asked me to give a talk to retired brothers at their mother house. I had assumed that like me, they had kept up with the changing world. Brother Joseph told me that they had all the facilities for my talk, including 'PowerPoint.' Just as in the old days I believed him. But no, they had not kept pace with the new Ireland. Their 'PowerPoint' was a single electric socket unevenly set in the wall. Theirs was a world still clinging to a clunking slide projector run from the socket.

With mixed emotions I watched them enter the lecture hall. One by one they shuffled in, pale, hunched, hair grey or totally gone. One walked with a stick, another with a walking frame. None of the brothers looked at me or at my projector but each made directly for a chair to which he seemed to have a particular attachment. Two came in wheelchairs, assisted by nurses who promptly left.

They gazed through me into the far distance, staring at something invisible to my eyes. There was not a twinkle of recognition: their minds were elsewhere, in a dream or lost in their bodies. Were these the brothers who stood before me in class, beguiling and inspiring me with their fine intellects? Were these the teachers who were intolerant of idiots, slow learners and sometimes of everything? I recognised none of them; their younger selves were hidden behind their wrinkles and sunken faces.

I tried to regale them with stories of nineteenth-century mapping. I tried, as some of them had done, to distil simplicity from the complicated. I told them about the limelight, that brilliant light that illuminated the darkness of mountain-tops when the fog and mist had cleared and had sufficient intensity to beam through the night.

But the limelight did not penetrate the fog in which they were shrouded. No eye flickered in understanding. Nor did the coloured maps of Waterford in 1841, showing the layout of the Norman walls and castles and towers, stimulate them. It was only then I realised that they too were mortal.

Behind the façade of the changing quay I could visualise the quay of my childhood map. I had often walked the riverside at night and stared at the matt black pontoons making holes in the shimmering waters that reflected the lights of the Quay. I loved to watch the *Rockabill,* in the black of the night, loading cattle bound for England and cranes hauling sacks from its holds and stacking them along the quayside. I wondered what was in them and I could vividly recall a schoolmate who drowned when his bicycle slipped between the ship and the wharf.

Further down, at the end of the Quay, was the *Great Western* with its two gangways, one for cattle, the other for people, both for export to Britain. There was constant shouting at the cattle gangway as men beat protesting and frightened cattle and drove them across a wide wooden bridge and through a big opening in the ship's side. Nearby, the ship's high deck was accessed by a long, narrow wooden gangplank. There was less commotion here and a more notice-ably silent emotion. Rows of people climbed the gangplank in resignation, heads bowed, shoulders drooped, brown cardboard cases in hand. Some, but only a few, had friends and relatives on hand to wave sad goodbyes. Many of the men, leaving wife and family behind to find work, would probably return in summer; some might never return, finding a new life in England instead. The women travelling hoped to find work and maybe a husband and a start to a better life.

Inscribed in my map, I had my own fictional ship at the top of Barrack Street. A lone telephone box like the bridge of a ship lit a small patch of the street. Five trees, the funnels of the ship, stood in a line behind it and, between them, concrete benches for the passengers. During the day the ship was deserted but it was a different story at night. Then, small groups, mostly women, stood around chatting and waiting. They were edgy, their talk strained; their faces had tense, worried and uncertain looks. They were waiting for the telephone to ring. This telephone, my father explained, was their only means of communication with their men or children who had taken the boat to England. A call had to be booked with the operator for a specific time and duration and the operator was sometimes not able to make contact if the cross-channel lines were busy or damaged. Or the telephone coin box might be jammed with a crooked coin or just full of coins.

Despite their expecting it, the ring of the phone always startled the women. Conversation stopped instantly and one would dash into the box and grab the receiver. A relieved smile might appear if the connection was made, a despairing look if the connection failed or if the call was not for them. After a successful call, news from England was shared by all. Faces beamed in delight if it was good and the conversation became animated. Perhaps a husband had secured a better-paid job and more money would be sent home or maybe he promised to come home for Christmas. A veil of tense concern covered the faces if the news was bleak; possibly a job was lost or it was hard to find work or lodgings. Each telephone conversation was discussed ad infinitum.

The image I retained of this place fascinated me and deepened as I grew up. The benches reminded me of tombs; they were made of solid concrete, about six feet long, nearly two feet wide and over two feet high. The trees, stunted because their branches were severely cut back every year, were like lives only partly lived. The phone box was a tenuous link between two worlds. It was also the keeper of the darker secrets that passed between those two places. The women were stuck, worried and anxious in one world, constantly restless and fearful of what might happen in the other.

Every Sunday morning the Barrack Street brass band exorcised this place. Their trumpets, trombones, tubas and drums played marching and rousing tunes, as if they were attempting to clear the air of the debris of the stresses and worries left behind by the waiting women each night.

I entered Michael Street, one of the streets my Little Match Girl had walked in the hope of a little kindness. What would she say now? Would passers-by give her a cent or more, or would they be too busy to notice her plight? Would she care that the developers were waiting to attack the street with yellow bulldozers and pull the heart from the city? I think she would have cared, for she was a girl of feeling. That was the destiny of Michael Street and the shops and houses that stretched back to Stephen Street, opposite my old school. The developers decided and the planners agreed that Waterford would be better off with a modern shopping mall complete with glass walls, air conditioning and piped music than a rattle bag of old shops and homes.

But at least the street was still there and I could wander through it with my Little Match Girl and our imaginary map. I could conjure up the stories my grandfather and my mother had told me. They were shopkeepers in 14 Michael Street. My grandmother was generous to all the little match girls and boys who lived in the Tanyard Arch and the lanes nearby. And they rewarded her every night by coming into her kitchen to kneel and recite the Rosary. Now, there were other match girls and match boys here; sitting in doorways, loitering with despairing, haunted, hungry, eyes. Please sir, could you spare twenty cent? Oh, please, to feed a hapless craving.

My mother once told me that in 1921 my unassuming grandfather put a gun on the counter of his drapery shop when the Black and Tans came to search the house. My grandmother was giving birth upstairs. He told them to get out or he would use it. Surprisingly, they left without a word. His had been a safe house where strangers on the run came and slept in my grandfather's tailor's shop in the attic before moving on again. No one ever asked who they were. What else came from that house? There was the music of his dance band, the Premier Dance Band, as it practised in the evenings. Passers-by stopped and listened and danced in the street to its renditions of Irvin Berlin's 'Blue Skies' or Hoagy Carmichael's 'Stardust'.

Who would remember the battle between the Blue Shirts and the Civic Guards in the 1930s if part of this street was swept away?

I doubted if anyone knew about the bottle that used to hold up my grandfather's window falling on the Blue Shirts as they marched up the street before it smashed on the ground, nearly causing a riot all of its own. Or what about the men and women who shopped for snuff across the road in my aunt's tobacco shop? I could see them now leaving with small newspaper cones filled with snuff. They could not wait to go home to pour it on to the back of their hands and take a sniff.

Now there was nothing there for anyone to see, only shops with transient tenants waiting until the bulldozers arrived. There were only the faded ghosts of those who once lived and worked there looking out forlornly from the blackness behind the upstairs windows. I would remember Michael Street as it was and honour its people and write down all I knew about them some day. The new maps of Waterford would tell nothing of their stories; they would instead show the new shopping mall, whenever it was built. There would be temporary tenants and multinational nameplates would adorn the shops. Would any of them leave a legacy of the stories of so many people?

How could I have forgotten the crow's foot directly opposite my bedroom window in my childhood home at 112 Cannon Street? A sapper had carved it into a green-brown stone of the convent wall sometime in the early 1840s. It watched over me for sixteen years. I was mesmerised by its perfection, its sharp-angled lines as clear as the day they were first carved. It was still there, as perfect as ever, waiting without hurry for someone else to stare at it from the same window. Four lines meeting at one point, silent, still, ever-present, beaming unspoken words through the glass pane of my bedroom window, filling the room with stories of the deeds of sappers.

This sapper might have mused about the heroics of his boss, Thomas Colby, climbing the most difficult mountains to support his men, such as his early-morning assault on Sawell in County Derry, risking his life in a violent storm that had not yet fully abated. The raging wind and rain had left a trail of destruction on the mountaintop: shelter, tents and equipment had been destroyed and the great theodolite barely saved. Thomas Colby appeared on the summit shortly after dawn and helped to restore shelter and equipment and recommence the observations. Maybe this was the sapper who carved the first benchmark on Poolbeg Lighthouse at low tide in April 1837. There must have been a story about that too. Perhaps I absorbed them all.

As I looked at it now it puzzled me that the sapper had placed his mark here. Why did he choose this nondescript stone in a wall made from thousands of stones, many of which were far more substantial? In any event he had carved a crow's foot a few yards away at the base of the cut limestone surround of the convent gate, a far more suitable and appropriate site. Two benchmarks in the space of a few yards were most unusual. Had the sapper rested with his back to the wall after he had completed the mark at the convent gate and looked across the dirt road at what was then Sullivan's field? There could have been nothing there but thorny ditches, green grass, dandelions, daisies and thistles. Had something from the heavens directed him to chisel the crow's foot into that stone, which would be perfectly visible from my bedroom window one hundred and twenty years hence? And was that mark the guiding star that led me from Cannon Street to Mountjoy House in the Phoenix Park? I liked to think that it was.

There were little memories that lingered on in other places too. A high, dark wall on a dimly-lit street on a black night. Three teenagers, two girls and a boy, the boy besotted by one of the girls. All happy on Summerhill, the music of the céilí still echoing in their ears. Unrestrained laughter at a trifle not now remembered. Disarmed, he looked into her eyes and she did not resist. But her father came and took her home and the cloak of awkward self-consciousness enveloped him again. A night remembered; a night never repeated.

As I was letting my roots sink back into the streets of Waterford I began to ponder again the three Thomases; Colby, Drummond and Larcom. I had set aside their mapping legacy and, of necessity, replaced it with maps of modern Ireland. But the roots of those modern maps lay with Colby and his companions. I could not let their achievements be forgotten. I decided that their maps, just like the history of the country, had to be made available to the people of Ireland – and not only those who would use them for academic research.

I began a new venture, to computerise Thomas Colby's entire legacy, all the maps he had produced, and make them available over the Web, to people in Ireland as well as the Irish scattered around the world. To seek out and computerise more than fifty thousand maps, all the editions up to 1913, when the first twenty-five-inch mapping series was completed, was a mammoth undertaking. We at the Ordnance Survey did not have the full archive. Much had passed to the National Archives. We trawled our own archives, the National Library and the library of the Royal Irish Academy to get the best copy of every map. It took three years to complete the computerisation of the six-inch and the twenty-five-inch series. We did more that just put individual maps on the Web. We joined all the adjacent map sheets together to form a seamless map of the country.

Now everyone could roam pre- and post-Famine Ireland and marvel at the changes to the landscape since the 1840s. The descendants of Irish emigrants could sit at their computer screens in America, Australia or Argentina and travel the country in search of their roots. They could visit the placenames their grandparents had spoken of and associate images of the landscape with them.

It was all completed by the end of 2005. The entire country had been remapped and now Thomas Colby's older maps sat side by side with them in the virtual world of cyberspace. It was time for me to depart. I left Ordnance Survey in mid-2006.

The Two Cartographers (Epilogue)

The enormousness of my spontaneous outburst in 1998 when I put my arms tightly around Eilís and said, 'I'm back. I'm back,' never faded from my mind. It had not struck me until that moment how very much I had been engrossed in a solitary world of work, travel and stress. Nor had it registered how much Eilís had supported me all that time and the extent to which I took that support for granted. She was always there, the rock of the family, our anchor, attending to everyone's needs, always the one to get things done, yet dedicated to her career as a teacher.

Years later I asked her how she managed through those years when my priority was the job of getting Ireland remapped. She made no answer but left the room. She returned some time later with an old, well-worn book and asked if I would like to hear her read something. She said, 'It's from a poem by John Donne called 'A Valediction: Forbidding Mourning':

> *If they be two, they are two so*
> *As stiff twin compasses are two:*
> *Thy soul, the fix'd foot, makes no show*
> *To move, but doth, if th' other do.*
>
> *And though it in the centre sit,*
> *Yet when the other far doth roam,*
> *It leans, and harkens after it,*
> *And grows erect, as that comes home.*

Such wilt thou be to me, who must
Like th' other foot, obliquely run;
Thy firmness makes my circle just,
And makes me end where I began.

We sat in silence; more words would have no meaning now. It was spring and a blackbird's song of renewal descended through the chimney and filled the room as it did each springtime. Ceaselessly it sang with joy and abandon.

Blue sky, sunshine, gentle breeze, blackbird on the branch;
Spinning world, earthquakes, volcanoes, hurricanes, chaos;
Blackbird, patient, alert, chattering, singing,
Waiting for its mate's response.

It dawned on me, as I sat there reflecting on John Donne's poem, that I had carried a symbol of our relationship for the many years I wandered the fields and climbed the mountains of Ireland. It was another type of compass, my surveyor's compass. It had a central pin, delicate but strong, that never ceased to balance the rotating arm that often swung wildly, not knowing which direction it was seeking.

Eilís never gave up on me. She enthusiastically supported me as I came to terms with the new way of life that unfolded for me after 1998. It was she who encouraged me to follow my instinct when I was wavering about training to become a Reiki master. It was Eilís who encouraged me with her 'can do' attitude when I protested that I could not possibly attend the creative writing workshop at Glenstal Abbey to which I had been invited. I had always insisted that I did not have the ability to string ten sentences together, let alone write stories or contemplate writing a book. But she reminded me of stories of my childhood I had often told her and produced a copybook into which she had written them.

She regarded it as one of life's unfolding chapters when I left Ordnance Survey. There, I could have rested on my laurels with status and security for a further eight years. She took it in her stride when I continued to be the other foot of the compass, obliquely running, and began a career as a part-time international mapping consultant.

I brought my Irish experience to Colombia, Mexico and the Ukraine. The national mapping agencies of these countries were beginning the journey we in Ireland had travelled and wished to learn from our approach to modernisation. They had skilled people and in many instances the World Bank or the European Union was funding their acquisition of technology similar to that used by our Ordnance Survey. When I visited Colombia, equipment that had been purchased months previously was still in boxes and it was not uncommon in some countries for equipment never to be assembled. Installation, training and maintenance budgets were often extremely limited as they had to come from scarce internal resources.

Colombia, like Mexico and the Ukraine, needed the expertise of others to show them how all the pieces of technology – satellites, computers, aerial-survey equipment and associated software – could be combined to set up a paperless map-production system. They needed fundamentally new ways of working, ways of which they had no experience and about which they were apprehensive. This meant breaking up discrete departments, little independent empires in themselves, to form new cohesive teams.

I was the missionary, enthusiastically preaching the new gospel of automation, explaining to them my experiences of the benefits of riding on the leading edge of technology, emphasising the importance of staying on the technology merry-go-round without taking undue risks. The practical example was Ordnance Survey in Ireland. I left them with the benefit of all I knew, hoping that they had learned something from my wisdom. I was always aware that they too might have archives of maps from bygone ages that were in danger of being lost. My advice always included ways by which they could let the old sit comfortably with the new. It was something that preyed on my mind as I talked to colleagues in the mapping organisations in those countries. I wondered if they too would go on a relentless pursuit of modernisation at the expense of their mapping heritage. Would it take a curious historian, centuries hence, to sift through remnants of old maps in an attempt to discover the origin of a lost mapping tradition? I hoped not. I had been lucky. That part of me that once delighted in tracing mental images on to old maps of Waterford had resurfaced before it was too late. It had opened Thomas Colby's legacy to the people of the world. But why, I kept asking myself, had another part of me, so driven to modernise the maps of Ireland, been intolerant of the past?

And as I mused on this I had to write about the two cartographers in me, each at odds with the other, to see if we could reconcile our separate identities and find a resolution.

THE TWO CARTOGRAPHERS

Light of the morning sun splashing on the map's white paper, brightness of the room, enthusiasm of the participants, sterility of the subject; turning living maps into lines in the computer. Mr K. had been responsible, encouraged it, got them hooked on it, stripping the clothes off nature. Hawthorn and blackthorn ditches reduced to lines, fields of grazing cattle called nothing but polygons; points where streams merged and waters chattered, excitedly bringing news from distant places, known only as nodes. Desecration in the name of progress – and Mr K. had not foreseen it. He called it modernisation. Here was the ultimate retribution for his work: they were butchering his map, although there were thousands more they could dissect.

Many years before, Mr K. and Larry had made this map: coaxed the information from the west Cork countryside, measured angles and distances; collected names – streams, laneways, mountains, lakes – names written through time in the memories of the people who lived there. Larry and the natural world conversed, communicated through touch of soil and boot, responded through his silence to the call of larks and gulls, listened to the whispering voices of the wind in the leaves and paid homage to the wildness in the anger of a storm. He felt the stoicism in the cliffs, by turns pounded by the sea and caressed by gentle tides. His breath resonated with the earth.

His drawings captured nature's essence. Mr K. was too young to notice: getting the job done was his priority.

Dazzling light of morning sun beaming into Larry's car, splashes of mud on the windscreen like black spots on the sun, sunbeams illuminating the haze of pipe smoke in the car. One contented man driving, drawing on his pipe; the other impatient to get the job finished.

'Mr K.' – for that was how he addressed people, lending gravitas and formality to all he said – 'Mr K., I told you, my hunch was right last night: a good landlady, a comfortable bed and a chunk of liver with the fry this morning. What more could a man want to fortify himself for the hard day ahead?'

The Ford Cortina, thirteen years into its life, the signs of its approaching end evident, rattled along the road. One wing was decaying with rust, a windscreen wiper lay limply out of its socket, the driver's door was at odds with the colour of the car and the steering wheel was tenuously connected to the driving shaft. The fabric of the seats reeked of years of pipe smoke, part of the car's persona. Larry had insisted on driving; he would not hear of Mr K. bringing his new Volkswagen. 'This car knows the roads, I just have to set it off in the right direction, and it doesn't matter if the briars or furze give it the odd scratch.'

'Time to get a new one, Larry. This banger won't last much longer. She will let you down some day in the middle of nowhere. Anyway, a new car should be no bother to you with all the expenses you get for travelling.'

'Mr K., don't talk to me about expenses. They barely keep body and soul together. Some nights I can't afford the price of a dinner and I rely on the generosity of a decent landlady for a cup of tea and a biscuit before these weary bones hit the sack.

'You're not that badly off, Larry,' replied Mr K. with a smile.

'I bought this car new, the first ever, and the debt nearly crippled me. There's a good five years still left in her with a little bit of tinkering and a few second-hand parts. She's been loyal. Never let me down. We know each other so well that I can smell when she needs a few parts or a drop of oil. Definitely not. No more new cars for me. I'd be broke forever.'

Larry, a man reaching into his middle fifties, had spent his career surveying the country, having learned his trade from his elders. Every county was etched in his mind, represented by the number of tortuous hills and mountains he had climbed, bogs that he had traversed and roads he had walked or driven. Every county had its special houses where decent women had taken pity and fed him at the end of a tiring day. He was not known for easily parting with money; in fact he never seemed to have any. His small, worn, leather wallet never shed more than a few pence to buy a box of matches. Mr K., the new broom, an eager new manager straight from college, saw things differently.

'Come off it, Larry. You have more expenses than anyone in this job; twelve months for however many years on the road and an allowance for every mile you go. The car costs you nothing. The digs are cheap and every landlady takes pity on you. You have to be loaded.'

'Nothing could be further from the truth, Mr K.,' replied Larry matter-of-factly.

'Anyway, Larry, let's get on with our business. It's a great day to get work done and if we're smart about it we'll climb two hills and get them out of the way. We'll be able to move on tomorrow.'

Larry, silent, drew slowly on his unlit pipe, pausing as if waiting for words of wisdom to rise from the tobacco. He turned into a narrow tarred road, wide enough for a single car, and slowed from the thirty miles an hour he had travelled on the main road.

'Mr K., you still have your youth but this ageing body of mine has been dragging itself up mountains and through bogs for more years than you have lived. Now I listen to it, treat it with respect, follow its wishes.' His voice carried his age, deliberate, ponderous, self-pitying. His physique was a contradiction to all he had said: six feet tall, erect, fresh face, hair only slightly faded in colour. Neither stiffness of joints nor middle-age spread afflicted him yet.

'Mr K., I'm afraid something's amiss.' His voice was uncertain as he slowed again and turned into a rutted boreen, manoeuvring the wayward steering wheel in an attempt to avoid the deepest potholes, while creeping brambles from the hedgerows kissed the car in welcome. 'This place has changed beyond recognition. Maybe they've moved away from the isolation of the hills and into the town.'

The bright red gate hanging from a whitewashed pillar was open and he drove through as if coming home. A tidy slated cottage, its door and two windows painted the same bright red, its walls whitewashed, was out of kilter with the overgrown laneway. The red and purple flowers of the fuchsia to the side of the house, like ballerinas standing on tiptoe, greeted them. A black collie inspected the car from the porch and lazily stood to greet his guests. His mistress, thin and greying, the years of living lodged as wrinkles in her face, inquisitively came to the door, drying her hands on her navy crossover apron.

'Larry! Hello, Larry. You're welcome. I haven't seen you for ages. I thought you must have moved on to some other county. You have a new man with you. Are you climbing again today? It's a great day for it anyhow.'

'Good morning, Mam. It surely is a powerful day. It must be nearly two years since I was back here. But what happened the boreen? It used to be so well kept.'

Her eyes darkened and met the stony ground. Words came slowly and from a distant place. 'God took him shortly after your last visit. It was quick in the end. I'm only slowly getting back to normal.'

Larry leaned forward and clasped her hand in both of his, and in words as slow as hers whispered, 'I'm sorry for your troubles. He was in great form when I saw him last. We had a glass of Powers together. It must be terribly lonely for you without him.'

'Sure it is. I didn't know how much I depended on him. Will you come in for a cup of tea before you make the climb? The kettle is on the boil.'

In subdued voice Larry responded, 'No thanks, not just now if you don't mind. We had best be on our way. It will take an hour or more to get to the top and we need to beat the weather before it closes in. It will be well into evening before we finish. We'll be bet at the end of it.'

Two blackbirds chattered enthusiastically from the fuchsia,

ignoring the sadness that hung in the yard. A white cat watched,
ready to pounce.

A bird bounced off the office window and fell to the ground.
Nobody cared.

'All hedges and fences will be represented by closed polygons.
There will be no exceptions. Every field corner will be classified as a
node and identified by a number; narrow streams will be depicted as
single lines…'

'And what about double ditches, how will they be represented?'
Enthusiastic voices around the table stripped nature to the bone.
No one dissented. One drew irregular lines on the white board,
explaining procedures to represent the fields in the computer.

The sun beat into the airless room, the windows of which
could not be opened. The furniture, new, fake, laminated, posing
as mahogany; filing cabinets holding the alchemical secrets
of the reduction of nature's bounty; a neat desk in the corner,
a barrier between its owner and all who entered; plastic trays
holding files. A coffee percolator full, its aroma seducing those
present to get another fix, another burst of caffeine to pulse the
enthusiasm for the job in hand. Mr K. listened impassively. He
had been through this many times but could no longer muster
enthusiasm. Black-and-red lines on a white board, flowcharts and
unintelligible scribbles did nothing to stimulate him any longer.

Larry sat on the stone beside the front door and slowly slipped each
foot in turn into a wellington boot. Mr K., with obvious impatience,
grabbed the orange box containing the theodolite and vaulted the
gate into the field. 'A man could cause himself permanent damage
doing things like that,' said Larry, as he lazily undid the wire that
held the gate closed and pulled it open over the tufts of grass. 'Easy
does it, it'll be a long day.'

'You know the woman of the house well. I'm surprised you
didn't accept the offer of tea. That's not your form, you never refuse
hospitality.' Mr K. walked two steps ahead of Larry, hoping his
mischievous talk would drag him on.

Larry paused, took the pipe full of tobacco from his pocket, inhaled and pensively compacted the tobacco with a small steel rod. 'Mr K., you can't go to the well too often. I think our luck will be in and there might be more than a cup of tea when we come back down. Everything comes with a little patience.'

Sunlight, blue skies overhead and the gentlest of breezes brought life to the countryside. Briars growing on the field boundaries stretched meandering branches to find other places to root, blackberries bulging, red berries turning black. Honey bees, stealing pollen, buzzed and busied themselves in flowers. Thrushes announced the visitors' arrival in animated song. Two seagulls glided on air currents.

Larry was part of it all. Silently leaning his body forward as if in homage to the earth, he pushed up the hill, setting each foot firmly into the soil. Regularly he paused, expanding his chest as he drew on the unlit pipe and absorbed all that was in the landscape around him. Only occasionally did he break the silence.

'Mr K., what a job to have on a day like this. Enjoy nature at its best. It makes a man feel good to be alive.'

Mr K. was edgy and wanted to gallop on. 'I hope that crew is on Seefin. Otherwise they might find themselves back permanently in the office. They were mighty sluggish shifting themselves from the digs this morning.'

Fields gave way to carpets of tingly purple heather interspersed with rocky paths made by mountain sheep and stony furrows left by winter streams. Cliffs and sea to the south came into view. The climb became steeper. Rustling heather thick on the ground made it more difficult to walk. Larry's pauses became more frequent.

'I think a bit of a rest is in order, Mr K.' Larry settled himself on a large rock and sucked from the unlit pipe once more. 'I don't suppose you have enough tea in that flask for a thirsty man to borrow a cup. The heat would take it out of you.' He was the picture of a contented man on that rock. His request betrayed the innocence of a child, the weariness of an ageing man. Below, the little cottage where Larry hoped for more than a cup of tea was a tiny patch on the landscape.

The airless office grew warmer and discussion a little more fraught as each individual tried to thrust his idea forward. Coats and jumpers discarded on chairs; dregs of coffee, stained plastic cups, cartons of chilled water from the plastic dispenser littered the table. Another diagram, more complicated now, to lay down the rules for drawing rivers and the bridges that crossed them. The meandering river, the straight, the wide, the narrow, all made to fit a series of straight lines or curves. There could be no exceptions; It was cold, clinical. John, a computer graduate, had never surveyed the countryside, which made it easier to act with precision and straitjacket the landscape into a map. Outside the window were ash saplings, their fading green leaves motionless in the sunshine. Sparrows darted from them to the ground in search of food. Nobody noticed. Mr K. sat impassively.

'Did you ever think of returning to the office yourself, Larry?' asked Mr K., irritated by the slow progress. 'You're getting on and this work isn't getting any easier, it's work for a younger man. Your experience would be invaluable there, especially with all those new graduates who need to be trained.'

Larry paused again, focused his eyes in contemplation on the tobacco in the pipe and slowly compacted it with his steel rod. A long slow suck on the pipe drew forth a calm response.

'Mr K., look at those seagulls, gliding on the air currents and diving into the sea for fish. They're natural, part of the natural balance here. Can you imagine them surviving on the Bog of Allen: no sea, no fish, no salt air, no cliffs to nest on? A balance in nature, Mr K., that's what we all become in this job, smiling in the sun, stoical in the rain and wind, resting in its moods, seduced by its brighter days. Take this from me and there is nothing left, nothing to give except the emptiness of a hollow shell. Mr K., tell me truly, what do you see here?'

'I see the top of that hill and a job waiting to be done and over there I see another hill to climb and more work to be done and before the month is out we will have finished and moved on to the next job. The scenery might be nice but the job's got to be done.'

Larry, silent, climbed slowly and with an easy weariness towards the top. The glassy sea stretching to the horizon was now in full view. Two boats stationary near the cliffs drew lobster pots. Seagulls congregated at a single spot on the sea and a black diver disappeared through the glass. Larry's pipe, firmly supported by his hand, was still unlit. A lark flew from the grass, soared to the heavens and sang in the still, shimmering air. She hovered, barely in sight, distracting the observers from her nest.

Mr K. sat the theodolite on the small concrete pillar on top of the hill, working with its three foot-screws until it was perfectly level. Carefully he peered through the telescope, searching out the tops of the distant hills.

'I hope those other crews have set up the targets and aren't dossing somewhere. I'll have them back at the office if they are,' he said. He looked again through the telescope and exclaimed, 'Three targets, Larry! That's all there are out of the five we need. This is ridiculous! They were told that we needed them by eleven. That's over an hour ago.'

'Mr K., I'm sure they will appear any time now. They can't be far away. There's a long journey between each of those hills and a steep climb. We have plenty of time. Those boys never let me down.'

'But the weather? Look, the clouds are building out to sea and won't take long to reach us. We'll end up getting soaked with nothing done and having to come back again.'

'I think it's time for a bite of lunch while we're waiting. Worrying never hurried them up.' He slid his single rasher sandwich, taken from the breakfast table, from his pocket and unwrapped the napkin. With a contented face, he chewed each morsel slowly, as if drawing from it the same wisdom he drew from the pipe. Not a word did he speak. His silence merged with the stillness of the lark's singing and the lonely echoing cry of a seagull near the cliffs.

'Does it matter what the rocks at the bottom of the cliff look like? We'll represent them all – flat, jagged, big or small – by the same symbol. They're only rocks.' Mr K.'s concentration had lapsed, overcome by the oppressiveness of the room. John's voice came like a drill in and out of Mr K.'s hearing. There was no dissent. Sandwiches encased in hard plastic had arrived and eager hands were releasing the entombed food from crackling cases. Munching mouths were too busy to engage with John. More coffee to wash down the bland food and recharge the adrenalin for an after-lunch assault on the remains of the map. Diagrams pasted to the walls: red, blue, black, unintelligible lines, squiggles, circles, scribbles. The tapestry of landscape was reduced to this.

'They're mighty sandwiches you have, Mr K. The landlady must have fancied you. I wonder if you could spare some for a poor body who brought only one and is still feeling a little peckish after the hard climb.'

Without a word, Mr K. slipped the sandwiches between them and let the stillness settle again. The lark, high above, sang and the seagull cried.

'I think we're right, Mr K. Everything's in position; the lads did their job. Your eyes are sharper than mine. Will you measure the angles?' Larry's voice, lively and enthusiastic, penetrated Mr K.'s consciousness and jolted him from his drowsiness. He was annoyed that he had not been vigilant.

Yes, there they there; there had been no need to worry. Mr K. rotated the telescope and measured each angle six times, degrees, minutes, seconds. Larry wrote the figures into his book, calling them out to make sure he recorded everything accurately. Three hours it took and they were on their way back down the hill, both contented with a successful day.

'Mr K., I wonder if our luck has run out? Not even the dog is here to greet us.'

'No harm, we can get on the road earlier and write up our field books back at the digs. We need to sort things out for the climb tomorrow and make sure that the other crews improve their timekeeping.'

'I'll knock, just in case, to say thanks to Mary.'

She was there, flustered, rubbing her hands on her navy crossover. 'Sorry, I didn't hear you coming. I was out in the kitchen washing up after baking the bread. Come in. Come along.'

'Mam, I must say there's a great smell of fresh cooking coming through the door. It would make any man hungry, not to mention a weary body coming off the hill. But Mr K. here is impatient to go so we'd best be on our way back to the digs. We'll get something in the town later. Thanks for minding the car.'

'I'll not hear of it. You can't go without a cup of tea at least. You must be starving after that climb and with little food all day. My father, God rest him, wouldn't have allowed it. Come in and honour his passing.'

'Well Mr K., I suppose a quick cup of tea won't do us any harm and it will keep us going for a while.'

Larry's eyes brightened, his satisfaction unconcealed at the sight of the table in the centre of this neat room set out for a feast. The table, rectangular, was covered by a bright oilcloth, four bare wooden chairs tucked under it. Willow-patterned delph was neatly set before three of the chairs. In the centre was a bowl with a dozen hard-boiled eggs, a plate of home-cooked ham carved thick and chunky, cheese generously sliced, tomatoes, a cake of home-made brown bread, partly cut in slices, the bread knife beside it ready for action, two dishes of butter, two pots of jam, salt, pepper and a stand for the teapot.

'Mam, your hospitality is as good as ever. You must have been baking all day. There's enough food here for an army.'

'I was hoping you would join me for the tea. It's lonely eating on your own. And imagine I almost missed you. What would I do with all this food?'

Outside, the thickening clouds that followed them from the hill laid siege to the house and filled the yard with a gloomy light. Only a pinch squeezed through the small windows. Leaping flames from the turf fire sent shadows dancing across the walls. Larry surveyed the table and rubbed his hands with delight.

'A hot cup of tea would have been appreciated but you tempt a weary traveller with such a mighty spread. It would be a shame to waste it.' Facing the fire with the image of the Sacred Heart watching him, he spread soft butter thickly on brown bread. The flames of the fire exaggerated the movements of his arm as it cast its shadow on to the wall.

'Come on Mr K., eat up, there's plenty of food. Larry, you never told me this man's first name?'

'Richard,' said Mr K. 'But Larry insists on calling everyone by their surname.'

'It's so nice to have visitors. I've been awful lonely since my father died. I'm the last in line, not a niece or nephew to come and visit. Just a few neighbours and even they are getting scarce. Two in the next parish died two months ago.' Her voice, lonely and soft, held the memories dear.

She reminisced about how her forebears lived here right back to her great-grandparents, worked the land, helped the neighbours, celebrated births and deaths and all the great feast days of the calendar. Her grandfather used to talk about the sappers who camped on the hill Mr K. and Larry had just climbed. The sappers often waited for a month for clear nights to see the beacons on other hills and read angles with huge theodolites. Some of the beacons were thirty miles away. He used to talk about the sapper's mark on top of the hill, a small triangle with a central hole carved into the rock. She was sure it was still there.

'Did you see it?' she said, as if it was important that they did. 'But my grandfather used to talk about another sapper's mark, a love child left behind with a broken-hearted mother when an English sapper left for another destination.'

She said her grandfather knew two love children but they had taken the boat to America when they were young adults. They had never returned.

'Times were different then. Nobody was in a hurry. There was all the time in the world to sit and talk and go at the pace nature intended. And, you know, everything got done just the same. My grandfather used to sit on the big stone outside — he called it the wisdom stone — chatting for hours to men who came to help or were just passing by.'

'Indeed you're right,' said Larry, drawing wisdom from his deliberate, slow chewing of the ham. 'Times have changed. We do in less than a day what it took at least a month to do then. Now people like Mr K. here are thinking they can do the same without ever coming here.'

'Well, you have to change with the times, keep up with advances. People cannot be living in the past, living off past glories. Move fast, get things done. That's what it is all about,' replied Mr K.

'I don't think we're the better for it though,' said Larry, watching the flames dance in the fire. 'I've spent my life travelling the countryside, getting to know its moods, its likes and dislikes. I have a feel and a respect for it. I get to know people like Mary here and we swap stories of times gone by and keep the memories of our ancestors and their times alive. That's what life is about. Work and money keep the body together; people and nature keep the soul alive. Do you not agree, Mr K.?'

'I tell you, in the next ten years there will be no such thing as having to climb mountains and wait for the weather. Computer technology and satellites will do it all. And it will be the same with everything. Time is money nowadays.'

'That would be a terrible tragedy.' Mary's lonely voice betrayed fear now. 'To let all the memories and all the stories of the past disappear. Life would be so awful, people would be so unhappy if money and time became the gods of the world.'

Mary could not imagine a life where she could not sit out on her grandfather's stone, let the world go by and have nature bring her its delights to feed and nourish her soul, just as earlier today when she watched the bees suck honey from the fuchsia, the cat asleep on the window sill, the blackbirds and the thrush singing in the bushes.

'My soul would shrivel up if I rushed from morning to night having to get things done. What would we do without time to sit and talk? What would people living alone do for company?'

'That's the way it will be, whether we like it or not,' Mr K. responded with impatience.

Night had closed in and the clouds had begun to spill. The fire and the Sacred Heart lamp gave a cosy light. 'Relics of the past,' Mr K. might have said. Shadows cast by the glow and the gentle flames moved nobly across the walls like visiting ancestors of this house, remembering times past. They might have told stories of the last time the sappers came by; perhaps even now conversed among themselves about the return of the new-age sappers.

Larry, seduced by the food, the warmth and comfort of the room and the kindred spirit this woman was, settled back in his chair, eased his pipe from his pocket and pressed the tobacco with his finger. He lit it and drew serenity from its stem. Mr K. was at ease too, bewitched by the shape-shifting shadows dancing on the wall. Strangely they had comforted him. Perhaps they had whispered words of wisdom into his soul.

Mary slowly stood and lifted three glasses and a bottle of Power's from the dresser. 'A little something for the night ahead,' she gently said. 'My father would have insisted.'

They drank to the memory of her father and as she watched Larry and Mr K. with the flickering shadows dancing around them, she thought the shadows had enveloped the two and made them one.

Bibliography

_____. *Ordnance Survey Letters, 1834-1841*. Dublin: Royal Irish Academy.

_____. *Ordnance Survey in Ireland: An Illustrated Record*. Dublin: Ordnance Survey of Ireland, 1991.

Andrews, J. H. *A Paper Landscape: The Ordnance Survey in Nineteenth-Century Ireland*. Oxford, 1975.

Clare, W. *A Young Irishman's Diary (1836-1847). Being Extracts from the Early Journal of John Keegan of Moate*. London, 1928.

Close, Charles. *The Early Years of the Ordnance Survey*. Institution of Royal Engineers, 1926: reprint Newton Abbot, 1969.

Colby, Col. *Ordnance Survey of the County of Londonderry, Volume the First: Memoir of the City and Northwest Liberties of Londonderry, Parish of Templemore*. Dublin, 1837.

Day, A. and P. McWilliams (eds.). *Ordnance Survey Memoirs of Ireland*. 40 vols. Belfast and Dublin, 1990-1998.

Barry O'Brien, R. *Thomas Drummond: Under-Secretary in Ireland, 1835-40, Life and Letters*. London, 1889; digital reprint, Kessinger.

Ferguson McLennan, John. *Memoir of Thomas Drummond: Under-Secretary to the Lord Lieutenant of Ireland, 1835-40*. Edinburgh, 1867.

Herity, M. (ed.). *Ordnance Survey Letters, Down*. Dublin: Four Masters Press, 2001.

James, H. *Methods and Processes Adopted for Production of the Maps of the Ordnance Survey*. London: HMSO, 1902.

Portlock, J.E. *Memoir of the Life of Major-General Colby: Together with a Sketch of the Origin and Progress of the Ordnance Survey of Great Britain*. London, 1869; digital reprint, Kessinger.